3rd International Conference on Hadron Physics 2011

(TROIA'11)

Journal of Physics: Conference Series Volume 348

Canakkale, Turkey
22-25 August 2011

Editors:

Güray Erkol
Ayşe Küçükarslan
Altuğ Özpineci

ISBN: 978-1-62276-694-9
ISSN: 1742-6588

Printed from e-media with permission by:

Curran Associates, Inc.
57 Morehouse Lane
Red Hook, NY 12571

Some format issues inherent in the e-media version may also appear in this print version.

Printed by Curran Associates, Inc. (2013)

Additional copies of this publication are available from:

Curran Associates, Inc.
57 Morehouse Lane
Red Hook, NY 12571 USA
Phone: 845-758-0400
Fax: 845-758-2634
Email: curran@proceedings.com
Web: www.proceedings.com

TABLE OF CONTENTS

PREFACE

The 3rd International Conference on Hadron Physics, TROIA'11 was held at Canakkale, Turkey on 22–25 August 2011. Ozyegin University, Middle East Technical University, Canakkale Onsekiz Mart University and HadronPhysics2 Consortium sponsored the conference. Its aim was to bring together the experts and young scientists working on experimental and theoretical hadron physics. About 60 participants from 12 countries attended the conference.

The topics covered included:

- Chiral Perturbation Theory
- QCD Sum Rules
- Effective Field Theory
- Exotic Hadrons
- Hadron Properties from Lattice QCD
- Experimental Results and Future Perspectives
- Hadronic Distribution Amplitudes

The conference presentations were organized such that the morning sessions contained invited talks and the afternoon sessions were devoted to contributed talks and poster presentations. The speakers of the invited talks were: D Melikhov, M Nielsen, M Oka, E Oset, S Scherer, T T Takahashi and R Wanke.

The conference venue was a resort hotel near Canakkale. As a social program, a guided full-day excursion to the excavation site of the ancient town of Troia and Assos was organized.

We believe that this conference provided a medium for young scientists and experts in the field to effectively communicate and share ideas. We would like to express our sincere thanks to all participants for their contributions and stimulating discussions. We are also grateful to the Scientific Secretary, Kadir Utku Can, and all other members of the Organizing Committee for their patience and efforts.

13 February 2012

The Editors
Güray Erkol
Ayşe Küçükarslan
Altuğ Özpineci

All papers published in this volume of *Journal of Physics: Conference Series* have been peer reviewed through processes administered by the proceedings Editors. Reviews were conducted by expert referees to the professional and scientific standards expected of a proceedings journal published by IOP Publishing.

Baryon chiral perturbation theory

S Scherer

Institut für Kernphysik, Johannes Gutenberg-Universität, D-55099 Mainz, Germany

Abstract. We provide a short introduction to the one-nucleon sector of chiral perturbation theory and address the issue of power counting and renormalization. We discuss the infrared regularization and the extended on-mass-shell scheme. Both allow for the inclusion of further degrees of freedom beyond pions and nucleons and the application to higher-loop calculations. As applications we consider the chiral expansion of the nucleon mass to order $\mathcal{O}(q^6)$ and the inclusion of vector and axial-vector mesons in the calculation of nucleon form factors. Finally, we address the complex-mass scheme for describing unstable particles in effective field theory.

1. Introduction

Effective field theory (EFT) is a powerful tool in the description of the strong interactions. Generally speaking, an EFT is a low-energy approximation to some underlying, more fundamental theory. The EFT is expressed in terms of a suitable set of effective degrees of freedom, dominating the phenomena in the low-energy region (see Table 1). In the context of the strong interactions, the underlying theory is quantum chromodynamics (QCD)—a gauge theory with color SU(3) as the gauge group. The fundamental degrees of freedom of QCD, quarks and gluons, carry non-zero color charge. Under normal conditions they do not appear as free particles. One assumes that any asymptotically observed hadron must be in a color-singlet state, i.e., a physically observable state is invariant under SU(3) color transformations. The strong increase of the running coupling for large distances possibly provides a mechanism for the color confinement. For the low-energy properties of the strong interactions another phenomenon is of vital importance. The masses of the up and down quarks and, to a lesser extent, also of the strange quark are sufficiently small that the dynamics of QCD in the chiral limit, i.e., for massless quarks, is believed to resemble that of the "real" world. Although a rigorous mathematical proof is not yet available, there are good reasons to assume that a dynamical spontaneous symmetry breaking (SSB) emerges from the chiral limit, i.e., the ground state of QCD has a lower symmetry than the QCD Lagrangian. For example, the comparatively small masses of the pseudoscalar octet, the absence of a parity doubling in the low-energy spectrum of hadrons, and a non-vanishing scalar singlet quark condensate are indications for SSB in QCD. According to the Goldstone theorem, a breakdown of the chiral $\mathrm{SU}(3)_L \times \mathrm{SU}(3)_R$ symmetry at the Lagrangian level to the $\mathrm{SU}(3)_V$ symmetry in the ground state implies the existence of eight massless pseudoscalar Goldstone bosons. The finite masses of the pseudoscalar octet of the real world are attributed to the explicit chiral symmetry breaking by the quark masses in the QCD Lagrangian. Both the vanishing of the Goldstone boson masses in the chiral limit and the vanishing interactions in the zero-energy limit provide a natural starting point for a derivative and quark-mass expansion. The corresponding EFT is called (mesonic) chiral perturbation

Journal of Physics: Conference Series **348** (2012) 012001 doi:10.1088/1742-6596/348/1/012001

	Fundamental theory	Effective field theory
Theoretical framework	QCD	ChPT
Degrees of freedom	Quarks and gluons	Goldstone bosons (+ other hadrons)
Parameters	g_3 + quark masses	Low-energy coupling constants + quark masses

Table 1. Comparison of QCD and ChPT.

theory (ChPT) [1, 2, 3], with the Goldstone bosons as the relevant effective degrees of freedom (see, e.g., Refs. [4, 5, 6, 7, 8] for an introduction and overview).

Mesonic ChPT may be extended to also include other hadronic degrees of freedom. The prerequisite for such an effective field theory program is (a) a knowledge of the most general effective Lagrangian and (b) an expansion scheme for observables in terms of a consistent power counting method. In the following we will outline some developments of the last decade in devising renormalization schemes leading to a simple and consistent power counting for the renormalized diagrams of baryon chiral perturbation theory (BChPT) [9].

2. Renormalization and Power Counting

2.1. Effective Lagrangian and power counting

The effective Lagrangian relevant to the one-nucleon sector consists of the sum of the purely mesonic and πN Lagrangians, respectively,

$$\mathcal{L}_{\text{eff}} = \mathcal{L}_\pi + \mathcal{L}_{\pi N} = \mathcal{L}_\pi^{(2)} + \mathcal{L}_\pi^{(4)} + \ldots + \mathcal{L}_{\pi N}^{(1)} + \mathcal{L}_{\pi N}^{(2)} + \ldots,$$

which are organized in a derivative and quark-mass expansion [1, 2, 9]. For example, the lowest-order basic Lagrangian $\mathcal{L}_{\pi N}^{(1)}$, already expressed in terms of renormalized parameters and fields, is given by

$$\mathcal{L}_{\pi N}^{(1)} = \bar{\Psi}\left(i\gamma_\mu\partial^\mu - m\right)\Psi - \frac{1}{2}\frac{\mathrm{g}_A}{F}\bar{\Psi}\gamma_\mu\gamma_5\tau^a\partial^\mu\pi^a\Psi + \ldots, \tag{1}$$

where m, g_A, and F denote the chiral limit of the physical nucleon mass, the axial-vector coupling constant, and the pion-decay constant, respectively. The ellipsis refers to terms containing external fields and higher powers of the pion fields. When studying higher orders in perturbation theory one encounters ultraviolet divergences. As a preliminary step, the loop integrals are regularized, typically by means of dimensional regularization. For example, the simplest dimensionally regularized integral relevant to ChPT is given by [4]

$$I(M^2,\mu^2,n) = \mu^{4-n}\int\frac{\mathrm{d}^n k}{(2\pi)^n}\frac{i}{k^2 - M^2 + i0^+} = \frac{M^2}{16\pi^2}\left(R + 2\ln\frac{M}{\mu}\right) + O(n-4),$$

where $R = \frac{2}{n-4} - [\ln(4\pi) + \Gamma'(1)] - 1$ approaches infinity as $n \to 4$. The 't Hooft parameter μ is responsible for the fact that the integral has the same dimension for arbitrary n. In the process of renormalization the counter terms are adjusted such that they absorb all the ultraviolet divergences occurring in the calculation of loop diagrams [10]. This will be possible, because we include in the Lagrangian all of the infinite number of interactions allowed by symmetries [11]. At the end the regularization is removed by taking the limit $n \to 4$. Moreover, when renormalizing, we still have the freedom of choosing a renormalization prescription. In this context the finite pieces of the renormalized couplings will be adjusted such that renormalized

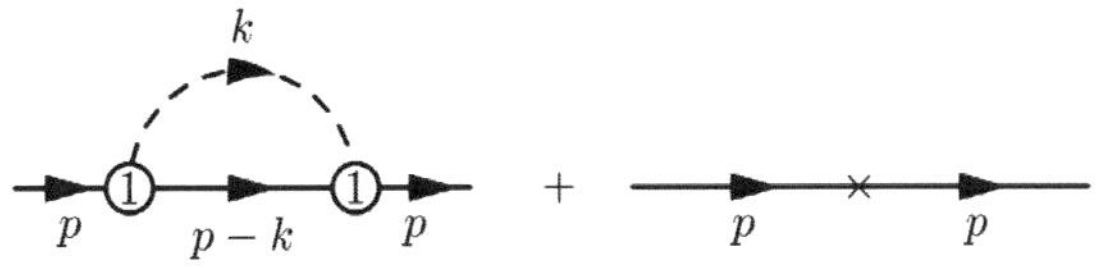

Figure 1. Renormalized one-loop self-energy diagram. The number 1 in the interaction blobs refers to $\mathcal{L}_{\pi N}^{(1)}$. The cross generically denotes counter-term contributions.

diagrams satisfy the following power counting: a loop integration in n dimensions counts as q^n, pion and fermion propagators count as q^{-2} and q^{-1}, respectively, vertices derived from $\mathcal{L}_{\pi}^{(2k)}$ and $\mathcal{L}_{\pi N}^{(k)}$ count as q^{2k} and q^k, respectively. Here, q collectively stands for a small quantity such as the pion mass, small external four-momenta of the pion, and small external three-momenta of the nucleon. The power counting does not uniquely fix the renormalization scheme, i.e., there are different renormalization schemes leading to the above specified power counting.

2.2. Example: One-loop contribution to the nucleon mass

In the mesonic sector, the combination of dimensional regularization and the modified minimal subtraction scheme $\widetilde{\text{MS}}$ leads to a straightforward correspondence between the chiral and loop expansions [2, 3]. By discussing the one-loop contribution of Fig. 1 to the nucleon self energy, we will see that this correspondence, at first sight, seems to be lost in the baryonic sector. According to the power counting specified above, after renormalization, we would like to have the order $D = n \cdot 1 - 2 \cdot 1 - 1 \cdot 1 + 1 \cdot 2 = n - 1$. An explicit calculation yields [8]

$$\Sigma_{\text{loop}} = -\frac{3\mathrm{g}_A^2}{4F^2}\left\{(\not{p}+m)I_N + M^2(\not{p}+m)I_{N\pi} - \frac{(p^2-m^2)\not{p}}{2p^2}[(p^2-m^2+M^2)I_{N\pi}+I_N-I_\pi]\right\},$$

where $M^2 = 2B\hat{m}$ is the lowest-order expression for the squared pion mass in terms of the low-energy coupling constant B and the average light-quark mass $\hat{m}$ [2]. The relevant loop integrals are defined as

$$I_\pi = \mu^{4-n}\int\frac{\mathrm{d}^n k}{(2\pi)^n}\frac{i}{k^2-M^2+i0^+}, \tag{2}$$

$$I_N = \mu^{4-n}\int\frac{\mathrm{d}^n k}{(2\pi)^n}\frac{i}{k^2-m^2+i0^+}, \tag{3}$$

$$I_{N\pi} = \mu^{4-n}\int\frac{\mathrm{d}^n k}{(2\pi)^n}\frac{i}{[(k-p)^2-m^2+i0^+]}\frac{1}{k^2-M^2+i0^+}. \tag{4}$$

The application of the $\widetilde{\text{MS}}$ renormalization scheme of ChPT [2, 9]—indicated by "r"—yields

$$\Sigma_{\text{loop}}^r = -\frac{3\mathrm{g}_{Ar}^2}{4F^2}\left[M^2(\not{p}+m)I_{N\pi}^r + \ldots\right].$$

The expansion of $I_{N\pi}^r$ is given by

$$I_{N\pi}^r = \frac{1}{16\pi^2}\left(-1+\frac{\pi M}{m}+\ldots\right),$$

resulting in $\Sigma_{\text{loop}}^r = \mathcal{O}(q^2)$. In other words, the $\widetilde{\text{MS}}$-renormalized result does not produce the desired low-energy behavior which, for a long time, was interpreted as the absence of a systematic power counting in the relativistic formulation of ChPT.

The expression for the nucleon mass m_N is obtained by solving the equation

$$m_N - m - \Sigma(m_N) = 0,$$

from which we obtain for the nucleon mass in the $\widetilde{\rm MS}$ scheme [9],

$$m_N = m - 4c_{1r}M^2 + \frac{3g_{Ar}^2 M^2}{32\pi^2 F_r^2}m - \frac{3g_{Ar}^2 M^3}{32\pi F_r^2}. \tag{5}$$

At $\mathcal{O}(q^2)$, Eq. (5) contains besides the undesired loop contribution proportional to M^2 the tree-level contribution $-4c_{1r}M^2$ from the next-to-leading-order Lagrangian $\mathcal{L}_{\pi N}^{(2)}$.

The solution to the power-counting problem is the observation that the term violating the power counting, namely, the third on the right-hand side of Eq. (5), is *analytic* in the quark mass and can thus be absorbed in counter terms. In addition to the $\widetilde{\rm MS}$ scheme we have to perform an additional *finite* renormalization. For that purpose we rewrite

$$c_{1r} = c_1 + \delta c_1, \quad \delta c_1 = \frac{3mg_A^2}{128\pi^2 F^2} + \ldots \tag{6}$$

in Eq. (5) which then gives the final result for the nucleon mass at $\mathcal{O}(q^3)$:

$$m_N = m - 4c_1 M^2 - \frac{3g_A^2 M^3}{32\pi F^2}. \tag{7}$$

We have thus seen that the validity of a power-counting scheme is intimately connected with a suitable renormalization condition. In the case of the nucleon mass, the $\widetilde{\rm MS}$ scheme alone does not suffice to bring about a consistent power counting.

2.3. Infrared regularization and extended on-mass-shell scheme

Several methods have been suggested to obtain a consistent power counting in a manifestly Lorentz-invariant approach. We will illustrate the underlying ideas in terms of a typical one-loop integral in the chiral limit,

$$H(p^2, m^2; n) = \int \frac{d^n k}{(2\pi)^n} \frac{i}{[(k-p)^2 - m^2 + i0^+][k^2 + i0^+]},$$

where $\Delta = (p^2 - m^2)/m^2 = \mathcal{O}(q)$ is a small quantity. Applying the dimensional counting analysis of Ref. [12], the result of the integration is of the form

$$H \sim F(n, \Delta) + \Delta^{n-3} G(n, \Delta),$$

where F and G are hypergeometric functions which are analytic for $|\Delta| < 1$ for any n.

In the infrared regularization of Becher and Leutwyler [13] one makes use of the Feynman parameterization

$$\frac{1}{ab} = \int_0^1 dz \frac{1}{[az + b(1-z)]^2}$$

with $a = (k-p)^2 - m^2 + i0^+$ and $b = k^2 + i0^+$. The resulting integral over the Feynman parameter z is then rewritten as

$$H = \int_0^1 dz \ldots = \int_0^\infty dz \ldots - \int_1^\infty dz \ldots,$$

where the first, so-called infrared (singular) integral satisfies the power counting, while the remainder violates power counting but turns out to be regular and can thus be absorbed in counter terms.

The central idea of the extended on-mass-shell (EOMS) scheme [14, 15] consists of subtracting those terms which violate the power counting as $n \to 4$. Since the terms violating the power counting are analytic in small quantities, they can be absorbed by counter-term contributions. In the present case, we want the renormalized integral to be of the order $D = n-1-2 = n-3$. To that end one first expands the integrand in small quantities and subtracts those integrated terms whose order is smaller than suggested by the power counting. The corresponding subtraction term reads

$$H^{\rm subtr} = \int \frac{d^n k}{(2\pi)^n} \frac{i}{[k^2 - 2p\cdot k + i0^+][k^2 + i0^+]}\bigg|_{p^2=m^2}$$

and the renormalized integral is written as $H^R = H - H^{\rm subtr} = \mathcal{O}(q)$ as $n \to 4$.

2.4. Remarks

Using a suitable renormalization condition, one obtains a consistent power counting in manifestly Lorentz-invariant baryon ChPT including, e.g., (axial) vector mesons [16] or the $\Delta(1232)$ resonance [17] as explicit degrees of freedom. The infrared regularization of Becher and Leutwyler [13] has been reformulated in a form analogous to the EOMS renormalization [18]. The application of both infrared and extended on-mass-shell renormalization schemes to multi-loop diagrams was explicitly demonstrated by means of a two-loop self-energy diagram [19]. A treatment of unstable particles such as the rho meson or the Roper resonance is possible in terms of the complex-mass scheme (CMS) [20, 21].

3. Applications

In the following we will illustrate a few selected applications of the manifestly Lorentz-invariant framework to the one-nucleon sector.

3.1. Nucleon mass to $\mathcal{O}(q^6)$

The nucleon mass m_N provides a good testing ground for applications of BChPT, as it does not depend on any momentum transfers and the chiral expansion therefore corresponds to an expansion in the quark masses. For this reason, the calculation of the nucleon mass has been performed in various renormalization schemes [9, 22, 23, 13, 15]. With the exception of Ref. [9], these schemes have in common that they establish the connection between the chiral and the loop expansions, analogous to the mesonic sector. A calculation that only includes one-loop diagrams can be performed up to and including $\mathcal{O}(q^4)$. The general form of the chiral expansion is given by

$$m_N = m + k_1 M^2 + k_2 M^3 + k_3 M^4 \ln\frac{M}{\mu} + k_4 M^4 + \ldots, \tag{8}$$

where $M^2 = 2B\hat{m}$ is the lowest-order expression for the squared pion mass, the ellipsis stands for higher-order terms, and μ is a renormalization scale. As an example, in the EOMS scheme the expressions for the k_i are given by [15]

$$\begin{aligned} k_1 &= -4c_1, \quad k_2 = -\frac{3g_A^2}{32\pi F^2}, \quad k_3 = \frac{3}{32\pi^2 F^2}\left(8c_1 - c_2 - 4c_3 - \frac{g_A^2}{m}\right), \\ k_4 &= \frac{3g_A^2}{32\pi^2 F^2 m}(1 + 4c_1 m) + \frac{3}{128\pi^2 F^2}c_2 - 2(8e_{38} + e_{115} + e_{116}). \end{aligned} \tag{9}$$

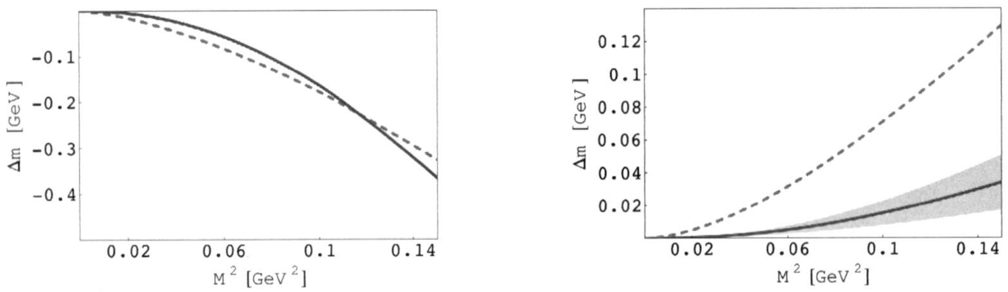

Figure 2. Pion-mass dependence of terms contributing to the chiral expansion of the nucleon mass. Left panel: The solid line corresponds to $k_5 M^5 \ln \frac{M}{\mu}$, the dashed line to $k_2 M^3$. Right panel: The solid line corresponds to $k_7 M^6 \ln^2 \frac{M}{\mu}$, the dashed line to $k_3 M^4 \ln \frac{M}{\mu}$. The grey band indicates an error estimate (see Ref. [27]).

Here, the c_i and e_j are low-energy constants (LECs) of the second- and fourth-order baryonic Lagrangians, respectively. The expression of Eq. (8) together with estimates of the various LECs [24] was used in Ref. [25] to determine the nucleon mass in the chiral limit,

$$\begin{aligned} m &= m_N - \Delta m \\ &= (938.3 - 74.8 + 15.3 + 4.7 + 1.6 - 2.3)\,\text{MeV} \\ &= 882.8\,\text{MeV}, \end{aligned} \tag{10}$$

i.e., $\Delta m = 55.5\,\text{MeV}$. Contributions to the nucleon mass at $\mathcal{O}(q^5)$, i.e., including two-loop diagrams, were first considered in Ref. [23], and a complete calculation to $\mathcal{O}(q^6)$ was performed in Refs. [26, 27]. The higher-order contributions take the form

$$\begin{aligned} m_N = m &+ k_1 M^2 + k_2 M^3 + k_3 M^4 \ln \frac{M}{\mu} + k_4 M^4 \\ &+ k_5 M^5 \ln \frac{M}{\mu} + k_6 M^5 + k_7 M^6 \ln^2 \frac{M}{\mu} + k_8 M^6 \ln \frac{M}{\mu} + k_9 M^6. \end{aligned} \tag{11}$$

Since various so-far undetermined LECs enter the expressions for some of the higher-order k_i it is not possible to give an accurate estimate of all terms in Eq. (11). However, the fifth-order contribution $k_5 M^5 \ln \frac{M}{\mu}$ is found to be $k_5 M^5 \ln \frac{M}{m_N} = -4.8\,\text{MeV}$ at the physical pion mass with $\mu = m_N$, while $k_6 M^5 = 3.7\,\text{MeV}$ or $k_6 M^5 = -7.6\,\text{MeV}$ depending on the choice of the third-order LEC d_{16} [27]. Equation (11) can also be used to examine the pion-mass dependence of the nucleon mass, which plays an important role in the extrapolation of lattice QCD to physical quark masses. Figure 2 shows the comparison of various terms in Eq. (11) as a function of the squared pion mass. While the right panel shows the expected suppression of the higher-order term over the whole pion-mass range, the left panel indicates that the term $k_5 M^5 \ln \frac{M}{\mu}$ becomes as large as $k_2 M^3$ for a pion mass of roughly $M \sim 360\,\text{MeV}$. While this is not a reliable prediction of the behavior of higher-order contributions since only the leading nonanalytic parts are considered, the pion mass range at which the power counting is no longer applicable agrees with the estimates found in Refs. [28, 29].

3.2. Probing the convergence of perturbative series

The issue of the convergence of perturbative calculations is presently of great interest in the context of chiral extrapolations of baryon properties (see, e.g., Refs. [30, 31, 32]). A possibility

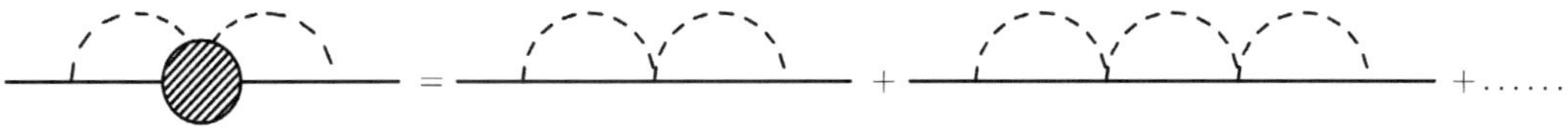

Figure 3. Iterated contribution to the nucleon self energy.

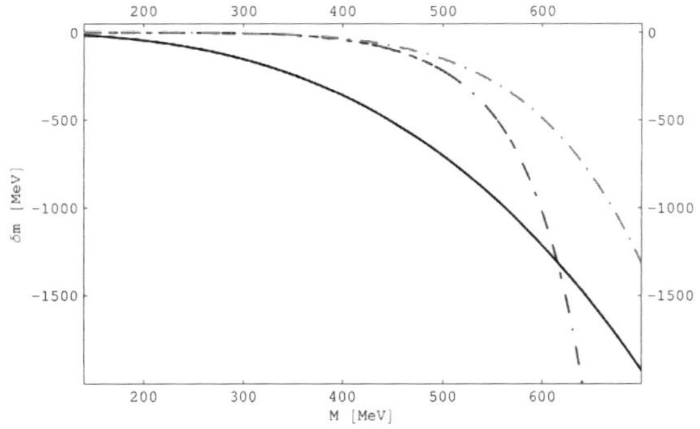

Figure 4. Contributions to the nucleon mass as functions of M. Solid line: $\mathcal{O}(q^3)$ contribution, dashed line: δm of Eq. (12); dashed-dotted line: two-loop diagram of Fig. 3.

of exploring the convergence of perturbative series consists of summing up certain sets of an infinite number of diagrams by solving integral equations exactly and comparing the solutions with the perturbative contributions [29]. Figure 3 shows a graphical representation of an iterated contribution to the nucleon self energy originating from the Weinberg-Tomozawa term in the πN scattering amplitude. The result is of the form [29]

$$\delta m = -\frac{3g_A^2}{4F^2}\frac{N}{D}, \qquad (12)$$

where N and D are closed expressions in terms of the loop functions of Eqs. (2) - (4). By expanding Eq. (12) in powers of $1/F^2$ one can identify the contributions of each diagram separately. Using the IR renormalization scheme and substituting $m = 883$ MeV, $m_N = 938.3$ MeV, $F = 92.4$ MeV, $g_A = 1.267$ and $M = 139.6$ MeV one obtains

$$\delta m = -0.00233530\,\mathrm{MeV} = (-0.00230219 - 0.00003305 - 0.00000007 + \ldots)\,\mathrm{MeV}. \qquad (13)$$

The first term in the perturbative expansion reproduces the non-perturbative result well and the higher-order corrections are clearly suppressed. Figure 4 shows δm of Eq. (12) together with the leading contribution (first diagram in Fig. 3) and the leading non-analytic correction to the nucleon mass $\delta m_3 = -3g_A^2 M^3/(32\pi F^2)$ [9] as functions of M. As can be seen from this figure, up to $M \sim 500$ MeV the non-perturbative sum of higher-order corrections is suppressed in comparison with the δm_3 term. Also, the leading higher-order contribution reproduces the non-perturbative result quite well. On the other hand, for $M \gtrsim 600$ MeV the higher-order contributions are no longer suppressed in comparison with δm_3.

3.3. Electromagnetic form factors

Imposing the relevant symmetries such as translational invariance, Lorentz covariance, the discrete symmetries, and current conservation, the nucleon matrix element of the electromagnetic

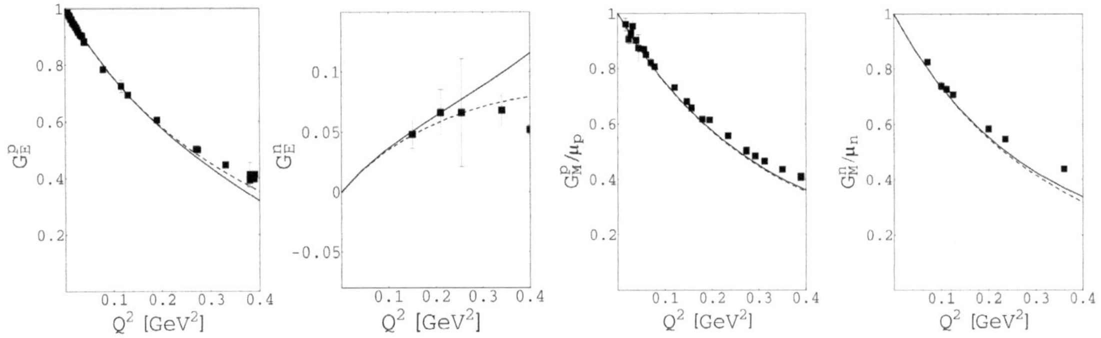

Figure 5. The Sachs form factors of the nucleon in manifestly Lorentz-invariant chiral perturbation theory at $\mathcal{O}(q^4)$ including vector mesons as explicit degrees of freedom. Full lines: results in the extended on-mass-shell scheme; dashed lines: results in infrared regularization. The experimental data are taken from Ref. [36].

current operator $J^\mu(x)$ can be parameterized in terms of two form factors,

$$\langle N(p')|J^\mu(0)|N(p)\rangle = \bar{u}(p')\left[F_1^N(Q^2)\gamma^\mu + i\frac{\sigma^{\mu\nu}q_\nu}{2m_p}F_2^N(Q^2)\right]u(p), \quad N=p,n, \tag{14}$$

where $q = p' - p$, $Q^2 = -q^2$, and m_p is the proton mass. At $Q^2 = 0$, the so-called Dirac and Pauli form factors F_1 and F_2 reduce to the charge and anomalous magnetic moment in units of the elementary charge and the nuclear magneton $e/(2m_p)$, respectively,

$$F_1^p(0) = 1, \quad F_1^n(0) = 0, \quad F_2^p(0) = 1.793, \quad F_2^n(0) = -1.913.$$

The Sachs form factors G_E and G_M are linear combinations of F_1 and F_2,

$$G_E^N(Q^2) = F_1^N(Q^2) - \frac{Q^2}{4m_p^2}F_2^N(Q^2), \quad G_M^N(Q^2) = F_1^N(Q^2) + F_2^N(Q^2), \quad N = p,n.$$

Calculations in Lorentz-invariant baryon ChPT up to fourth order fail to describe the proton and nucleon form factors for momentum transfers beyond $Q^2 \sim 0.1\,\mathrm{GeV}^2$ [33, 34]. In Ref. [33] it was shown that the inclusion of vector mesons can result in the re-summation of important higher-order contributions. In Ref. [35] the electromagnetic form factors of the nucleon up to fourth order have been calculated in manifestly Lorentz-invariant ChPT with vector mesons as explicit degrees of freedom. A systematic power counting for the renormalized diagrams has been implemented using both the extended on-mass-shell renormalization scheme and the reformulated version of infrared regularization. The inclusion of vector mesons results in a considerably improved description of the form factors (see Fig. 5). The most dominant contributions come from tree-level diagrams, while loop corrections with internal vector meson lines are small [35].

3.4. Axial and induced pseudoscalar form factors

Assuming isospin symmetry, the most general parametrization of the isovector axial-vector current evaluated between one-nucleon states is given by

$$\langle N(p')|A^{\mu,a}(0)|N(p)\rangle = \bar{u}(p')\left[\gamma^\mu\gamma_5 G_A(Q^2) + \frac{q^\mu}{2m_N}\gamma_5 G_P(Q^2)\right]\frac{\tau^a}{2}u(p), \tag{15}$$

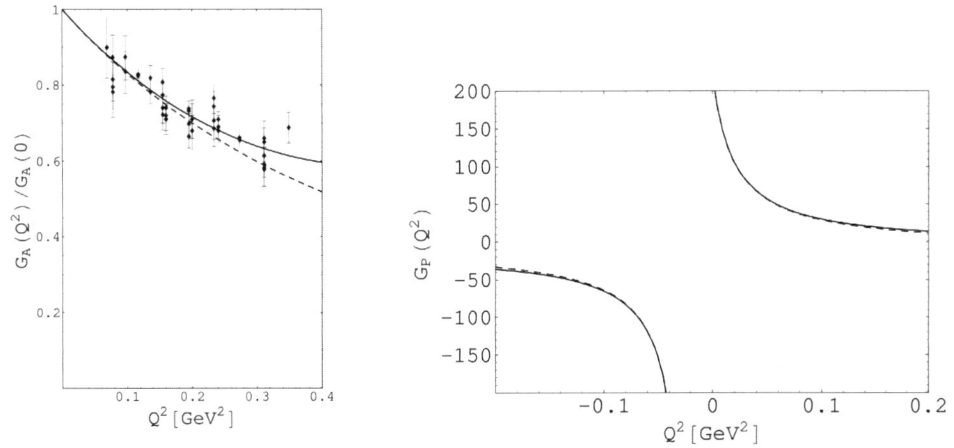

Figure 6. Left panel: Axial form factor G_A in manifestly Lorentz-invariant ChPT at $\mathcal{O}(q^4)$ including the axial-vector meson $a_1(1260)$ explicitly. Full line: result in infrared renormalization, dashed line: dipole parametrization. The experimental data are taken from Ref. [38]. Right panel: The induced pseudoscalar form factor G_P at $\mathcal{O}(q^4)$ including the axial-vector meson $a_1(1260)$ explicitly. Full line: result with axial-vector meson; dashed line: result without axial-vector meson. One can clearly see the dominant pion pole contribution at $Q^2 \approx -M_\pi^2$.

where $q = p' - p$, $Q^2 = -q^2$, and m_N denotes the nucleon mass. $G_A(Q^2)$ is called the axial form factor and $G_P(Q^2)$ is the induced pseudoscalar form factor. The value of the axial form factor at zero momentum transfer is defined as the axial-vector coupling constant, $g_A = G_A(Q^2 = 0) = 1.2694(28)$, and is quite precisely determined from neutron beta decay. The Q^2 dependence of the axial form factor can be obtained either through neutrino scattering or pion electroproduction. The second method makes use of the so-called Adler-Gilman relation [37] which provides a chiral Ward identity establishing a connection between charged pion electroproduction at threshold and the isovector axial-vector current evaluated between single-nucleon states (see, e.g., Ref. [38, 39] for more details). The induced pseudoscalar form factor $G_P(Q^2)$ has been investigated in ordinary and radiative muon capture as well as pion electroproduction (see Ref. [40] for a review).

In Ref. [41] the form factors G_A and G_P have been calculated in BChPT up to and including order $\mathcal{O}(q^4)$. In addition to the standard treatment including the nucleon and pions, the axial-vector meson $a_1(1260)$ has also been considered as an explicit degree of freedom. The inclusion of the axial-vector meson effectively results in one additional low-energy coupling constant which has been determined by a fit to the data for $G_A(Q^2)$. The inclusion of the axial-vector meson results in an improved description of the experimental data for G_A (see Fig. 6), while the contribution to G_P is small.

4. Complex-mass scheme and effective field theory

In Sec. 3.3 we saw how the inclusion of virtual vector mesons generates an improved description of the electromagnetic form factors, for which ordinary chiral perturbation theory does not produce sufficient curvature. So far the inclusion of virtual vector mesons has been restricted to low-energy processes in which the vector mesons cannot be generated explicitly. However, one would also like to investigate the properties of hadronic resonances such as their masses and widths as well as their electromagnetic properties. An extension of chiral effective field theory to the momentum region near the complex pole corresponding to the vector mesons was proposed in Ref. [20], in which the power-counting problem was addressed by applying the complex-mass scheme (CMS) [42, 43, 44, 45] to the effective field theory. Since the ρ mass is not treated as a small quantity, the presence of large external four-momenta, e.g., in terms of the zeroth component, leads to a considerable complication regarding the power counting of loop diagrams. To assign a chiral order to a given diagram it is first necessary to investigate all possibilities how the external momenta could flow through the internal lines of that diagram. Next, when assigning powers to propagators and vertices, one needs to determine the chiral order for a given flow of external momenta. Finally, the smallest order resulting from the various assignments is defined as the chiral order of the given diagram. The application of the CMS to the renormalization of loop diagrams amounts to splitting the bare parameters of the Lagrangian into renormalized parameters and counter terms and choosing the renormalized masses as the *complex* poles of the dressed propagators in the chiral limit, $M_R^2 = (M_\chi - i\Gamma_\chi/2)^2$. The result for the chiral expansion of the pole mass and the width of the ρ meson to $\mathcal{O}(q^4)$ reads [20]

$$
\begin{aligned}
M_\rho^2 &= M_\chi^2 + c_x M^2 - \frac{g_{\omega\rho\pi}^2 M^3 M_\chi}{24\pi} \\
&\quad + \frac{M^4}{32\pi^2 F^2}\left(3 - 2\ln\frac{M^2}{M_\chi^2}\right) + \frac{g_{\omega\rho\pi}^2 M^4}{32\pi^2}\left(1 - \ln\frac{M^2}{M_\chi^2}\right), && (16)\\
\Gamma &= \Gamma_\chi + \frac{\Gamma_\chi^3}{8M_\chi^2} - \frac{c_x \Gamma_\chi M^2}{2M_\chi^2} - \frac{g_{\omega\rho\pi}^2 M^3 \Gamma_\chi}{48\pi\, M_\chi} + \frac{M^4}{16\,\pi\, F^2 M_\chi}. && (17)
\end{aligned}
$$

Here, M^2 is the lowest-order expression for the squared pion mass, F the pion-decay constant in the chiral limit, c_x a low-energy coupling constant of the $\pi\rho$ Lagrangian, and $g_{\omega\rho\pi}$ a coupling constant. The nonanalytic terms of Eq. (16) agree with the results of Ref. [46]. Both mass M_χ and width Γ_χ in the chiral limit are input parameters in this approach. The numerical importance of the different contributions has been estimated using

$$F = 0.092\,\text{GeV}, \quad M = 0.139\,\text{GeV}, \quad g_{\omega\rho\pi} = 16\,\text{GeV}^{-1}, \quad M_\chi \approx M_\rho = 0.78\,\text{GeV},$$

resulting in the expansion (units of GeV^2 and GeV, respectively)

$$
\begin{aligned}
M_\rho^2 &= M_\chi^2 + 0.019\,c_x - 0.0071 + 0.0014 + 0.0013\,,\\
\Gamma &\approx \Gamma_\chi + 0.21\,\Gamma_\chi^3 - 0.016\,c_x\Gamma_\chi - 0.0058\,\Gamma_\chi + 0.0011\,. && (18)
\end{aligned}
$$

For pion masses larger than $M_\rho/2$ the ρ meson becomes a stable particle. For such values of the pion mass the series of Eq. (17) should diverge. Along similar lines, Ref. [21] contains a calculation of the mass and the width of the Roper resonance using the CMS.

5. Conclusion

In the baryonic sector new renormalization conditions have reconciled the manifestly Lorentz-invariant approach with the standard power counting. We have discussed some results of a two-loop calculation of the nucleon mass. The inclusion of vector and axial-vector mesons as explicit degrees of freedom leads to an improved phenomenological description of the electromagnetic and axial form factors, respectively. Work on the application to electromagnetic processes such as Compton scattering and pion production is in progress. When describing resonances in perturbation theory, one needs to take their finite widths into account. The CMS has opened a new window for describing unstable particles in EFT with a consistent power counting.

Acknowledgments
This work was made possible by the financial support from the Deutsche Forschungsgemeinschaft (SFB 443 and SCHE 459/2-1).

References
[1] S. Weinberg, Physica A **96**, 327 (1979).
[2] J. Gasser and H. Leutwyler, Annals Phys. **158**, 142 (1984).
[3] J. Gasser and H. Leutwyler, Nucl. Phys. **B250**, 465 (1985).
[4] S. Scherer, Adv. Nucl. Phys. **27**, 277 (2003).
[5] J. Bijnens, Prog. Part. Nucl. Phys. **58**, 521 (2007).
[6] V. Bernard, Prog. Part. Nucl. Phys. **60**, 82 (2008).
[7] S. Scherer, Prog. Part. Nucl. Phys. **64**, 1 (2010).
[8] S. Scherer and M. R. Schindler, Lect. Notes Phys. **830**, 1 (2012).
[9] J. Gasser, M. E. Sainio, and A. Švarc, Nucl. Phys. **B307**, 779 (1988).
[10] J. C. Collins, *Renormalization*, Cambridge University Press, Cambridge, 1984.
[11] S. Weinberg, *The Quantum Theory of Fields. Vol. 1: Foundations*, Cambridge University Press, Cambridge, 1995.
[12] J. Gegelia, G. S. Japaridze, and K. S. Turashvili, Theor. Math. Phys. **101**, 1313 (1994) [Teor. Mat. Fiz. **101**, 225 (1994)].
[13] T. Becher and H. Leutwyler, Eur. Phys. J. C **9**, 643 (1999).
[14] J. Gegelia and G. Japaridze, Phys. Rev. D **60**, 114038 (1999).
[15] T. Fuchs, J. Gegelia, G. Japaridze, and S. Scherer, Phys. Rev. D **68**, 056005 (2003).
[16] T. Fuchs, M. R. Schindler, J. Gegelia, and S. Scherer, Phys. Lett. B **575**, 11 (2003).
[17] C. Hacker, N. Wies, J. Gegelia, and S. Scherer, Phys. Rev. C **72**, 055203 (2005).
[18] M. R. Schindler, J. Gegelia, and S. Scherer, Phys. Lett. B **586**, 258 (2004).
[19] M. R. Schindler, J. Gegelia, and S. Scherer, Nucl. Phys. **B682**, 367 (2004).
[20] D. Djukanovic, J. Gegelia, A. Keller, and S. Scherer, Phys. Lett. B **680**, 235 (2009).
[21] D. Djukanovic, J. Gegelia, and S. Scherer, Phys. Lett. B **690**, 123 (2010).
[22] V. Bernard, N. Kaiser, J. Kambor, and U.-G. Meißner, Nucl. Phys. **B388**, 315 (1992).
[23] J. A. McGovern and M. C. Birse, Phys. Lett. B **446**, 300 (1999).
[24] T. Becher and H. Leutwyler, JHEP **0106**, 017 (2001).
[25] T. Fuchs, J. Gegelia, and S. Scherer, Eur. Phys. J. A **19**, 35 (2004).
[26] M. R. Schindler, D. Djukanovic, J. Gegelia, and S. Scherer, Phys. Lett. B **649**, 390 (2007).
[27] M. R. Schindler, D. Djukanovic, J. Gegelia, and S. Scherer, Nucl. Phys. **A803**, 68 (2008).
[28] U.-G. Meißner, PoS LAT2005, 009-1 (2006).
[29] D. Djukanovic, J. Gegelia, and S. Scherer, Eur. Phys. J. A **29**, 337 (2006).
[30] D. B. Leinweber, A. W. Thomas, and R. D. Young, Phys. Rev. Lett. **92**, 242002 (2004).
[31] M. Procura, T. R. Hemmert, and W. Weise, Phys. Rev. D **69**, 034505 (2004).
[32] S. R. Beane, Nucl. Phys. **B695**, 192 (2004).
[33] B. Kubis and U.-G. Meißner, Nucl. Phys. **A679**, 698 (2001).
[34] T. Fuchs, J. Gegelia and S. Scherer, J. Phys. G **30**, 1407 (2004).
[35] M. R. Schindler, J. Gegelia, and S. Scherer, Eur. Phys. J. A **26**, 1 (2005).
[36] J. Friedrich and Th. Walcher, Eur. Phys. J. A **17**, 607 (2003).
[37] S. L. Adler and F. J. Gilman, Phys. Rev. **152**, 1460 (1966).
[38] V. Bernard, L. Elouadrhiri, and U.-G. Meißner, J. Phys. G **28**, R1 (2002).
[39] T. Fuchs and S. Scherer, Phys. Rev. C **68**, 055501 (2003).
[40] T. Gorringe and H. W. Fearing, Rev. Mod. Phys. **76**, 31 (2004).
[41] M. R. Schindler, T. Fuchs, J. Gegelia, and S. Scherer, Phys. Rev. C **75**, 025202 (2007).
[42] R. G. Stuart, in Z^0 *Physics*, ed. J. Tran Thanh Van (Editions Frontiers, Gif-sur-Yvette, 1990), p. 41.
[43] A. Denner, S. Dittmaier, M. Roth, and D. Wackeroth, Nucl. Phys. **B560**, 33 (1999).
[44] A. Denner, S. Dittmaier, M. Roth, and L. H. Wieders, Nucl. Phys. **B724**, 247 (2005).
[45] A. Denner and S. Dittmaier, Nucl. Phys. Proc. Suppl. **160**, 22 (2006).
[46] D. B. Leinweber, A. W. Thomas, K. Tsushima, and S. V. Wright, Phys. Rev. D **64**, 094502 (2001).

Simulating a CP-violating topological term in gauge theories

Haralambos Panagopoulos [a*] **and Ettore Vicari** [b]
[a]Department of Physics, University of Cyprus, Lefkosia, CY-1678, Cyprus
[b]Dipartimento di Fisica, Università di Pisa and INFN, I-56127 Pisa, Italy
[*]Presented the talk

E-mail: haris@ucy.ac.cy, vicari@df.unipi.it

Abstract. We present recent results on the θ-dependence of four-dimensional SU(N) gauge theories, where θ is the coefficient of the CP-violating topological term in the Lagrangian. In particular, we study the scaling behavior of these theories, by Monte Carlo simulations at imaginary θ. The numerical results provide good evidence of scaling in the continuum limit. The imaginary θ dependence of the ground-state energy turns out to be well described by the first few terms of related expansions around $\theta = 0$, providing accurate estimates of the first few coefficients, up to $O(\theta^6)$.

Four-dimensional $SU(N)$ gauge theories have a nontrivial dependence on the parameter θ which appears in the Euclidean Lagrangian as

$$\mathcal{L}_\theta = (1/4)F^a_{\mu\nu}(x)F^a_{\mu\nu}(x) - i\,\theta\, q(x), \qquad q(x) = g^2/(64\pi^2)\,\epsilon_{\mu\nu\rho\sigma}F^a_{\mu\nu}(x)F^a_{\rho\sigma}(x), \tag{1}$$

where $q(x)$ is the topological charge density. The ground-state energy density $F(\theta)$ behaves as

$$\mathcal{F}(\theta) \equiv F(\theta) - F(0) = (1/2)\,\chi\theta^2 s(\theta), \tag{2}$$

where χ is the topological susceptibility at $\theta = 0$,

$$\chi \equiv \textstyle\int d^4x \langle q(x)q(0)\rangle_{\theta=0} = \langle Q^2\rangle_{\theta=0}/V, \quad Q \equiv \int d^4x\, q(x), \tag{3}$$

V is the spacetime volume and $s(\theta)$ is a dimensionless even function of θ such that $s(0) = 1$. Assuming analyticity at $\theta = 0$, $s(\theta)$ can be expanded as: $s(\theta) = 1 + b_2\theta^2 + b_4\theta^4 + \cdots$, where only even powers of θ appear. Large-N scaling arguments applied to the Lagrangian (1) indicate that the relevant scaling variable in the large-N limit is $\bar{\theta} \equiv \theta/N$. This implies that in this limit $\chi = O(1)$, while the coefficients b_{2i} are suppressed by powers of N, i.e. $b_{2i} = O(N^{-2i})$.

Due to the nonperturbative nature of the θ dependence, quantitative assessments have largely focused on the lattice formulation of the theory, using Monte Carlo (MC) simulations. However, the complex nature of the θ term in the Euclidean Lagrangian prohibits a direct MC simulation at $\theta \neq 0$. Information on the θ dependence of physically relevant quantities, such as the ground state energy and the spectrum, has been obtained by computing the coefficients of the corresponding expansions around $\theta = 0$. The coefficients of $s(\theta)$ can be determined from appropriate zero-momentum correlation functions of $q(x)$ at $\theta = 0$, which are related to the moments of the $\theta = 0$ probability distribution $P(Q)$ of the topological charge Q. Indeed

$$b_2 = -\frac{1}{12\,\chi V}\left[\langle Q^4\rangle - 3\langle Q^2\rangle^2\right]_{\theta=0}, \quad b_4 = -\frac{1}{360\,\chi V}\left[\langle Q^6\rangle - 15\langle Q^2\rangle\langle Q^4\rangle + 30\langle Q^2\rangle^3\right]_{\theta=0}, \tag{4}$$

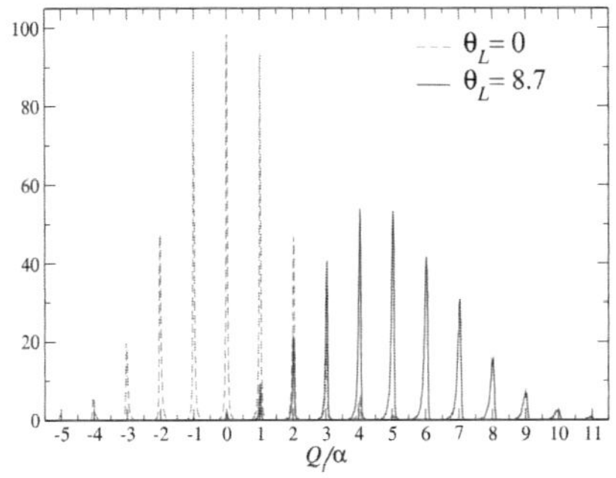

Figure 1. Distribution of the ratio Q_t/α, for $\beta = 6.2$ configurations at $\theta_L = 0$ and $\theta_L = 8.7$ ($\theta_i \approx 1.5$).

etc. They parameterize the deviations of $P(Q)$ from a simple Gaussian behavior. It has been shown that correlations involving multiple insertions of the topological charge can be defined in a nonambiguous, regularization independent way, and therefore b_{2i} are well defined renormalization group invariant quantities. The numerical evidence for a nontrivial θ-dependence, obtained through MC simulations, appears quite robust. We refer the reader to Ref. [1] for a recent review. On the other hand, MC simulations at $\theta = 0$ have only made it possible to estimate the ground-state energy up to $O(\theta^4)$. The large-N prediction $b_2 = O(N^{-2})$ has been already supported by numerical results; the calculation of the higher-order terms would provide a further check of the power-law suppression predicted by large-N arguments.

In this paper we consider imaginary values of θ, which make the Euclidean Lagrangian (1) real, thus making MC simulations possible. For further details and references, the reader may consult our Ref. [2]. Assuming analyticity at $\theta = 0$, the results provide quantitative information on the expansion around $\theta = 0$. Indeed, fits of the data to polynomials of imaginary θ may provide more accurate estimates of the coefficients, overcoming the rapid increase of statistical errors observed at $\theta = 0$. Perturbative renormalization-group (RG) arguments indicate that θ is a RG invariant parameter of the theory, thus the continuum limit should be approached while keeping θ fixed to any complex value. We find that this is indeed supported by the numerical data for the 4D SU(3) lattice gauge theory, at least for $|\theta| < \pi$, which are well described by the first few nontrivial terms of the expansion around $\theta = 0$.

Introducing the real parameter θ_i, defined by $\theta \equiv -i\theta_i$, Eq. (2) leads to:

$$\frac{\langle Q\rangle_{\theta_i}}{V} = -\frac{\partial \mathcal{F}(-i\theta_i)}{\partial \theta_i} = \chi\theta_i\left(1 - 2b_2\theta_i^2 + 3b_4\theta_i^4 + \cdots\right), \tag{5}$$

$$\frac{\langle Q^2\rangle^c_{\theta_i}}{V} \equiv \frac{\langle Q^2\rangle_{\theta_i} - \langle Q\rangle^2_{\theta_i}}{V} = -\frac{\partial^2 \mathcal{F}(-i\theta_i)}{\partial \theta_i^2} = \chi\left(1 - 6b_2\theta_i^2 + 15b_4\theta_i^4 + \cdots\right). \tag{6}$$

The nonperturbative formulation of the above theory on the lattice requires a discretization of the action, $S_L - \theta_L Q_L$; for S_L we use the plaquette gluon action, while for Q_L we employ the "twisted double plaquette" operator q_L ($Q_L = \sum_x q_L(x)$). Notice that this is not the only possible choice for q_L; the only requirement is that it have the correct continuum limit when $a \to 0$ (a: lattice spacing). In the continuum limit $q_L(x)$, being a local operator, behaves as

$$q_L(x) \longrightarrow a^4 Z_q\, q(x) + O(a^6), \tag{7}$$

where Z_q is a finite function of the bare coupling g_0, going to one in the limit $\beta \equiv 2N/g_0^2 \to \infty$. Thus, we have the correspondence: $\theta_i = Z_q\theta_L$, apart from $O(a^2)$ corrections. The renormalization Z_q may be evaluated by MC simulation at $\theta = 0$, computing

$$Z_q = \langle QQ_L\rangle_{\theta=0}/\langle Q^2\rangle_{\theta=0}\,, \tag{8}$$

where Q is an estimator such as those obtained by the overlap method or the cooling method, which are not affected by renormalizations, nor by nonphysical contact terms. Thus, the ratios

$$\langle Q\rangle_{\theta_i}/\langle Q^2\rangle_{\theta=0} = \theta_i - 2b_2\theta_i^3 + 3b_4\theta_i^5 + \ldots, \qquad \langle Q^2\rangle^c_{\theta_i}/\langle Q^2\rangle_{\theta=0} = 1 - 6b_2\theta_i^2 + 15b_4\theta_i^4 + \ldots, \tag{9}$$

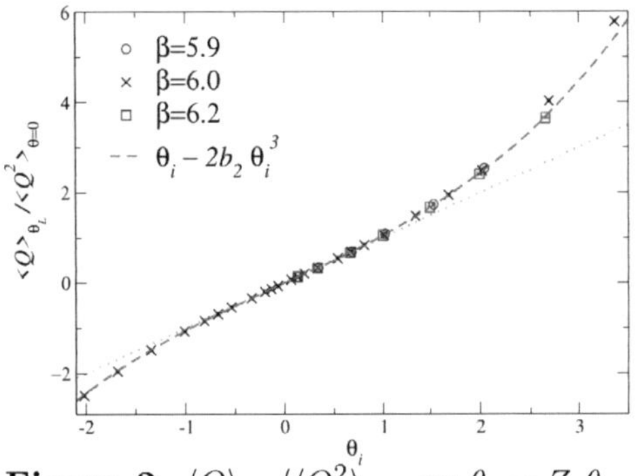

Figure 2. $\langle Q\rangle_{\theta_L}/\langle Q^2\rangle_{\theta=0}$ vs $\theta_i = Z_q\theta_L$.

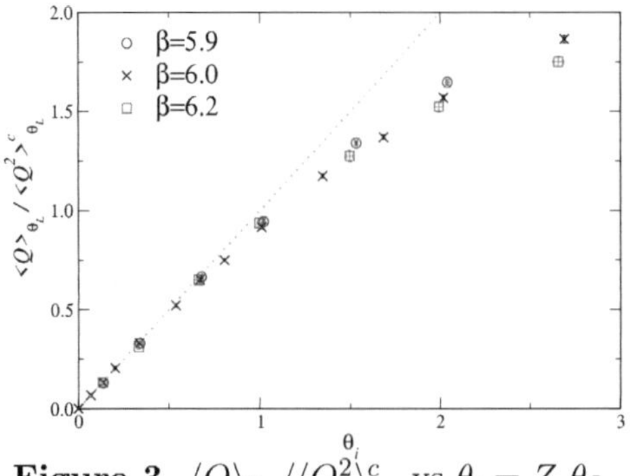

Figure 3. $\langle Q\rangle_{\theta_L}/\langle Q^2\rangle^c_{\theta_L}$ vs $\theta_i = Z_q\theta_L$.

are expected to have a well defined continuum limit as functions of θ_i.

We have carried out MC simulations of the 4D SU(3) lattice gauge theory, at $\beta = 5.9$, 6, 6.2, for lattice sizes $L = 16$, 16, 20, respectively; the simulations are carried out both at $\theta_L = 0$ and $\theta_L \neq 0$, within the region $|\theta_i| \lesssim \pi$. Since our numerical study requires high-statistics MC simulations, we choose the cooling method as estimator of the topological charge Q. The topological charge has been measured on cooled configurations (by locally minimizing the lattice action), using the twisted double plaquette operator. As is well known, this procedure leads to values $Q_t \simeq k\alpha$, where k is an integer and $\alpha \lesssim 1$. Once we determine α, we assign to Q the integer closest to Q_t/α. This cooling method for estimating Q, though less rigorous than the significantly more expensive overlap method, produces results in good agreement with it.

Fig. 1 shows the distributions of the ratio Q_t/α of $\beta = 6.2$ cooled configurations at $\theta_L = 0$ and $\theta_L = 8.7$ ($\theta_i = Z_q\theta_L \approx 1.5$). We note that these distributions cluster around integer values, also for rather large values of Q_t/α, both for $\theta_L = 0$ and $\theta_L = 8.7$.

MC simulations at $\theta = 0$ were performed at $\beta = 5.9$, 6.0, 6.2. Over 40 million sweeps per value of β were produced. The results for χ, b_2, b_4 and Z_q are reported in Ref. [2]. Providing our improved estimate of high-order coefficients, such as b_4, turned out to be very hard in $\theta = 0$ MC simulations, requiring huge statistics. The results for b_4 are consistent with zero, suggesting the bound $|b_4| \lesssim 0.005$, which is improved in $\theta \neq 0$ runs. Our estimates of Z_q (e.g. $Z_q(\beta = 6.0) = 0.135(1)$) reduce the uncertainty on Z_q as produced by other methods.

MC simulations at $\theta_L \neq 0$ are slower by approximately a factor of three, due to the complexity of the action. In $\theta_L \neq 0$ runs, ~ 3 million sweeps were produced for each value of β and θ_L. Figs. 2 and 3 show results for the ratios $\langle Q\rangle_{\theta_L}/\langle Q^2\rangle_{\theta=0}$ and $\langle Q\rangle_{\theta_L}/\langle Q^2\rangle^c_{\theta_L}$, versus $\theta_i = Z_q\theta_L$ (cf. Eq. (9)). The MC data at different β values follow the same curve, providing evidence of scaling. Scaling corrections, expected to be $O(a^2)$, are quite small, and tend to increase with increasing θ_i. This good scaling behavior corroborates the existence of a nontrivial continuum limit for any value of θ_i. Fitting our data to Eqs. (5,6,9) improves significantly the $\theta = 0$ results. In particular, a much smaller bound on b_4 is obtained: $|b_4| < 0.001$; also, we find: $b_2 = -0.026(3)$, which is clearly more precise than the estimate obtained from $\theta = 0$ runs only: $b_2 = -0.029(7)$.

Besides allowing more precise determinations of the θ expansion coefficients of the ground-state energy and other observables, using imaginary θ values might turn out useful in overcoming the dramatic critical slowing down of topological modes, by performing parallel tempering simulations with a set of θ values including θ=0; this is an *exact* MC algorithm for the model.

References

[1] E. Vicari and H. Panagopoulos, *θ dependence of SU(N) gauge theories in the presence of a topological term*, Phys. Rep. 470 (2009) 93 [`arXiv:0803.1593 hep-th`].

[2] H. Panagopoulos and E. Vicari, *The 4D SU(3) gauge theory with an imaginary θ term*, JHEP 11 (2011) 119 [`arXiv:1109.6815`].

Journal of Physics: Conference Series **348** (2012) 012003 doi:10.1088/1742-6596/348/1/012003

Kaonic atoms measurements at the DAΦNE accelerator: the SIDDHARTA experiment

A Rizzo[1], M Bazzi[1], G Beer[2], C Berucci[1], L Bombelli[3], A M Bragadireanu[4], M Cargnelli[5], C Curceanu (Petrascu)[1], G Corradi[1], A dUffizi[1], C Fiorini[3], T Frizzi[3], F Ghio[6], C Guaraldo[1], R S Hayano[7], M Iliescu[1], T Ishiwatari[5], M Iwasaki[8], P Kienle[9], P Levi Sandri[1], A Longoni[3], V Lucherini[1], J Marton[5], S Okada[1], D Pietreanu[4], T Ponta[4], A Romero Vidal[1], E Sbardella[1], A Scordo[1], H Shi[7], D L Sirghi[1,4], F Sirghi[1,4], H Tatsuno[7], A Tudorache[4], V Tudorache[4], O Vazquez Doce[1], E Widmann[5], B Wunschek[5], J Zmeskal[5]

[1] Laboratori Nazionali di Frascati - INFN, Via Enrico Fermi 40, Frascati 00044 (Rome), Italy
[2] Department of Physics and Astronomy, University of Victoria - Victoria B.C., Canada
[3] Politecnico di Milano, Sezione di Elettronica - Milano, Italy
[4] IFIN-HH - Magurele, Bucharest, Romania
[5] Stefan-Meyer-Institut fur subatomare Physik - Vienna, Austria
[6] INFN, Sezione di Roma 1 and Istituto Superiore di Sanita - Rome, Italy
[7] University of Tokyo - Tokyo, Japan
[8] RIKEN, The Institute of Physics and Chemistry Research - Saitama, Japan
[9] Physik Department, Technische Universitat Munchen - Garching, Germany

E-mail: `alessandro.rizzo@lnf.infn.it`

Abstract. Kaonic Hydrogen and Helium X-ray measurements play nowadays a fundamental role in testing the reliability of the Chiral Perturbation Theory as a realisation of the Quantum Chromodynamics at low energies. Dictated by both electromagnetic and strong interaction, X-ray transitions at lower energy levels of these complex bound systems offer indeed the unique opportunity to perform a threshold measurements of zero-energy meson-nucleon scattering. Nowadays the SIDDHARTA experiment at DAΦNE collider is the only apparatus which can provide such kind of measurements with the high precision needed to disentangle different theoretical calculation scenarios. In this work we present the SIDDHARTA experiment performances and results, with a focus on the main topics of light kaonic atom physics.

1. Introduction

The light kaonic atoms, matter of study of the SIDDHARTA experiment at the DAΦNE collider, are bound systems made of a light target atom and a K^- meson captured in an external atomic orbital, with a lifetime of the order of $10^{-12}s$. The difference between the atomic binding energies of these light systems (in the keV range) and the hadronic scale (about 1GeV) make them a very useful tool to perform experiments equivalent to scattering at vanishing energies: this allows, in the strangeness sector, to check the nature of the three flavor Chiral Symmetry Breaking through a real threshold measurement.

The experimental challenges of this kind of physics are strictly related to four main processes that occur during the lifetime of these bound systems: the capture, the formation, the cascade and the absorption process (see ref.[1] for an exhaustive description). After the capture of the K^- and the formation of the bound system, the meson starts a series of radiative transitions towards the lower lying levels (cascade process): in the last ones, when the meson is nearest to the nucleus, an X-ray with the energyemission dictated by both the electromagnetic and strong interaction occurs, bringing us information about the strong force contribution at very low energy. A high precision spectroscopic X-ray measurement of the lines related to these transitions allows to evaluate their shift (ϵ) and width (Γ) respect to the pure electromagnetic case. Recognising these parameters (ϵ,Γ) respectively as the real and the imaginary part of a complex scattering length through the Deser-Truemann formula [2], we can relate them to the effective chiral lagrangian using the Bethe-Salpeter equation [3]. From the experimental point of view the capture and the cascade process are the most important. The capture probability is maximum when the K^- kinetic energy is comparable to the kinetic energy of the atomic electron in the capturing orbital (13.6 eV for the Kaonic Hydrogen (KH)), a very low-energy monochromatic K^- beam is needed. The cascade process is the most important one for the spectroscopic measurement itself: after the depletion of all low lying atomic electrons till a formation of an hydrogenoid system nucleus - K^-, in the last part of the cascade, namely the Particle-Nucleus interaction regime (when the meson is nearest to the nucleus) , the X-ray transitions of experimental interest occur [1]. Moreover during this process several processes compete, causing the loss of the signal: the most important are the coulombian de-excitation, the elastic scattering and the Stark mixing, that increase with density (see ref.[4]). In the next section the SIDDHARTA apparatus is described, with a focus on how it can satisfy these requirements.

2. The SIDDHARTA experiment at the DAΦNE collider

The ideal low-energy monochromatic K^- beam to produce kaonic atoms in the target of the SIDDHARTA experiment is provided by the DAΦNE e^+e^- collider of the Laboratori Nazionali di Frascati, tuned on the Φ resonance at rest (1020 MeV). With a branching ratio of 49.2% the Φ decades in K^+K^- emitted back to back, with a momentum of 127 MeV/c and a spread $\Delta p/p$ of 0.1%. These particular characteristics offer several advantages for the SIDDHARTA experiment: first, the low kaons momentum allows to use a thin plastic degrader to slow them without enhance the energy spread, second the particular topology of the Φ decay allows to implement a trigger system to disentangle the kaons and the MIPs (Minimum Ionizing Particles). Moreover, a new collisional scheme implemented on the DAΦNE collider enhance the machine luminosity to $5 \times 10^{32} cm^{-2}s^{-1}$ through the Crab-Waist collisions, accounting nowadays the DAΦNE collider the ideal machine to perform this kind of physics [5]. To detect the particular topology of the Φ decay, the SIDDHARTA trigger system is constituted by two scintillators, placed immediately below and above the Interaction Point (IP). Through a TOF measurement, it allows to disentangle kaons from synchronous background with a separation of about 1 ns. In order to correlate the passage of a produced kaon with an X-ray emission coming from a kaonic atom, to reduce the background in the spectroscopic measurement, a very fast detector for high precision spectroscopy is needed. The Silicon Drift Detectors (SDD) combine a fast response with a good energy resolution (FWHM=150eV@5.9KeV): 144 SDDs with 1 cm^2 of active area surround the SIDDHARTA target, to detect the X-rays [6]. They were continuously monitored during the data taking with *in-situ* calibrations, one every 4 hours, using a sandwich of Ti and Cu foils and an X-ray tube. The produced K^-, passing through the scintillator of the trigger system and the plastic degrader, enters in the SIDDHARTA target cell, placed in the vacuum chamber and surrounded by the SDDs. To enhance the mean free path of the kaonic atoms in the target cell minimising the contribution of the competitor processes, SIDDHARTA uses a gaseus target. The temperature and the pressure of the target are carefully set, because the capture probability is proportional to the density of the target itself. In particular, for the different SIDDHARTA targets (kaonic H, ^{4}He (K^{4}He) and ^{3}He (K^{3}He)), the temperature and the pressure were set as follow: 23 K and 10^5 Pa for the KH, 20K and 10^5 Pa for the K^{3}He and 27K and 10^5 Pa for the ^{4}He. For a detailed description of the SIDDHARTA apparatus see ref.[7]. In the next section we present the SIDDHARTA results on these different light kaonic atoms.

3. SIDDHARTA results

- *Kaonic Hydrogen* (ref.[8]) : A set of data for a total amount of 340 pb^{-1} has been collected and analysed. The result for the shift ϵ and the width Γ for the transition K_α is: $\epsilon_{1s} = -283 \pm 36(stat) \pm 6(syst)eV$ and $\Gamma_{1s} = 541 \pm 89(stat) \pm 22(syst)eV$. Nowadays it is the best result available in literature.
- K^4He (ref.[7]): A set of data for a total amount of 20 pb^{-1} has been collected and analysed. A source of ^{55}Fe has been used during the data taking to check the systematics. The results for the shift ϵ for the transition L_α is: $\epsilon = 0 \pm 6(stat) \pm 2(syst)$. This is the first result obtained using a gaseous ^{4}He target, that confirm the result obtained in a previous experiment with a liquid target (see ref. [10]).
- K^3He (ref.[9]): A set of data for a total amount of 15 pb^{-1} has been collected and analysed. The results for the shift ϵ for the transition L_α is: $\epsilon = -1.7 \pm 2.7(stat) \pm 4(syst)$. This is the first measurement in the world for this light kaonic atom.

4. Conclusions and outlooks

Thanks to the monochromatic low-energy kaons beam delivered by the DAΦNE collider, wich allowed the choice of a gaseous target and the use of very-fast and high-precision detector for spectroscopy (SDDs), the SIDDHARTA experiment obtained a unique-quality results on light kaonic atoms. Presently an upgrade of the SIDDHARTA apparatus is ongoing for the SIDDHARTA-2 phase. The aim of this project is to study an enlarged scientific case, in particular the Lyman series of kaonic deuterium (Kd) and of K^3He and K^4He, never done before. This will allow to have an overall view of the strong interaction at low-energy in the strangeness sector in system with one, two, three and four nucleons.

We Acknowledge the support of the European Community-Research Infrastructure Integrating Activity Study of Strongly Interacting Matter (acronym HadronPhysics2, Grant Agreement no. 227431) under the Seventh Framework Programme of EU.

References

[1] Gotta D 2004 *Progress in Particle and Nuclear Physics* **52** 133
[2] Truemann T L *et al.* 1961 *Nucl. Phys.* **26** 57
[3] Itzykson C Zuber J 1980 *Quantum Field Theory* (New York, McGraw-Hill)
[4] Jensen T S *et al.* 2002 *Eur. Phys. J.* D **19** 165
[5] Boscolo M *et al.* 2010 *Nucl. Instr. and Meth.* A **621** 121
[6] SIDDHARTA Collaboration, Marton J *In the Proceedings of International Symposium on Detector Development for Particle, Astroparticle and Synchrotron Radiation Experiments (SNIC 2006), Menlo Park, California, 3-6 Apr 2006, pp 0196.*
[7] SIDDHARTA Collaboration, Bazzi M *et al.* 2009 *Phys. Lett.* B **681** 310
[8] SIDDHARTA Collaboration, Bazzi M *et al.* 2011 *Phys. Lett.* B **704** 113
[9] SIDDHARTA Collaboration, Bazzi M *et al.* 2011 *Phys. Lett.* B **697** 199
[10] Okada S *et al.* 2007 *Eur. Phys. J.* B **653** 387

Journal of Physics: Conference Series **348** (2012) 012004 doi:10.1088/1742-6596/348/1/012004

Vector meson-vector meson interaction and dynamically generated resonances

E Oset[1], L S Geng[2] and R Molina[1]

[1]Departamento de Física Teórica and IFIC, Universidad de Valencia-CSIC, E-46071 Valencia, Spain

[2]School of physics and nuclear energy engineering, Beihang University, Beijing 100191, China

E-mail: oset@ific.uv.es

Abstract. We report upon 11 composite meson states, dynamically generated from the vector meson-vector meson interaction using the local hidden gauge formalism within a unitary approach. Six of these states are associated to the $f_0(1370)$, $f_0(1710)$, $f_2(1270)$, $f_2'(1525)$, $a_2(1320)$ and $K_2^*(1430)$ resonances. At the same time we predict five other states with the quantum numbers of h_1, a_0, b_1, K_0^*, and K_1 which could be tested by future experiments.

1. Introduction

The chiral unitary approach, combining coupled-channels, unitarity and chiral Lagrangians, has been quite successful to deal with hadron interactions at low energies. The interaction leads to scattering matrices which have poles in the complex plane associated to resonances which are called dynamically generated (see Ref. [1] for an introduction and earlier references). Because these states are dynamically generated from meson-meson or meson-baryon interactions, they are viewed as composite particles or hadronic molecules, instead of genuine $q\bar{q}$ or qqq systems. Studies of their behavior in the large N_c limit have confirmed that they are largely meson-meson or meson-baryon composite systems, though some of them may contain a small genuine $q\bar{q}$ or qqq component.

The success of the chiral unitary approach has inspired a further extension to study the interaction between two vector mesons and between one vector meson and one baryon [2, 3, 4, 5, 6, 7]. Here, instead of using interaction kernels provided by ChPT, one uses transition amplitudes provided by the hidden-gauge Lagrangians, which lead to a suitable description of the interaction of vector mesons among themselves and of vector mesons with other mesons or baryons. Coupled-channel unitarity works in the same way as in the unitary chiral approach, but now the dynamics is provided by the hidden-gauge Lagrangians [8]. The use of this approach in several recent works [2, 3, 4, 5, 6, 7] has proved rather sucessful.

In this talk, we report on a recent study of vector meson-vector meson interaction using the unitary approach, briefly discussing the formalism and showing how 11 states are dynamically generated [3].

2. Formalism

We briefly outline the main ingredients of the unitary approach (see Refs. [2, 3] for more details). There are two basic building-blocks in this approach: transition amplitudes provided

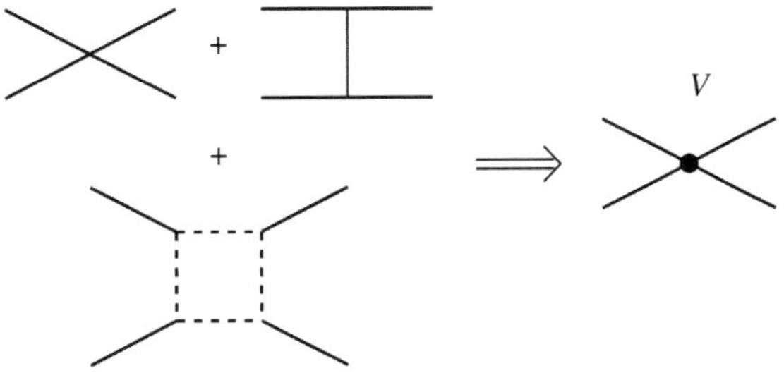

Figure 1. Transition amplitudes V appearing in the coupled-channel Bethe-Salpeter equation.

by the hidden-gauge Lagrangians [8] and a unitarization procedure. We adopt the Bethe-Salpeter equation method $T = (1 - VG)^{-1}V$ to unitarize the transition amplitudes V for s-wave interactions, where G is a diagonal matrix of the vector meson-vector meson one-loop function.

In Refs. [2, 3] three mechanisms, as shown in Fig. 1, have been taken into account for the calculation of the transition amplitudes V: the four-vector-contact term, the t(u)-channel vector-exchange amplitude, and the direct box amplitude with two intermediate pseudoscalar mesons. Other possible mechanisms, e.g. crossed box amplitudes and box amplitudes involving anomalous couplings, have been neglected, since their contributions have been shown to be small in the case of rho-rho scattering in Ref. [2].

Among the three mechanisms considered for V, the four-vector-contact term and $t(u)$-channel vector-exchange are responsible for the formation of resonances or bound states if the interaction generated by them is strong enough. In this sense, the dynamically generated states can be thought of as "vector meson-vector meson molecules." On the other hand, the consideration of the imaginary part of the direct box amplitude allows the generated states to decay into two pseudoscalars. It should be stressed that in the present approach these two mechanisms play quite different roles: the four-vector-contact term and the $t(u)$-channel vector-exchange one are responsible for generating the resonances whereas the direct box amplitude mainly contributes to their decays.

Since the energy regions we are interested in are close to the two vector-meson threshold, the three-momenta of the external vector mesons are smaller than the corresponding masses, $|\vec{q}|^2/M^2 \ll 1$, and therefore can be safely neglected. This considerably simplifies the calculation of the four-vector-contact term and the $t(u)$-channel vector-exchange amplitude, whose explicit expressions can be found in Appendix A of Ref. [3]. To regularize the box diagram, one takes the empirical form factors used in the study of vector-meson decays [9].

To regularize the two vector-meson one-loop function one has to introduce either cutoffs in the cutoff method or subtraction constants in the dimensional regularization method. Further details concerning the values of these parameters can be found in Ref. [3]. The results presented in this talk are obtained using the dimensional regularization method with the values of the subtraction constants given in Ref. [3]. In particular, we have fine-tuned the values of three subtraction constants to reproduce the masses of the three tensor states, i.e., the $f_2(1270)$, the $f_2'(1525)$, and the $K_2^*(1430)$. The masses of the other eight states are predictions.

3. Results and discussions

By looking at poles of the scattering matrix T on the second Riemann sheet, we find 11 states in nine strangeness-isospin-spin channels as shown in Table 1. Theoretical masses and widths are obtained with two different methods: In the "pole position" method, the mass corresponds to the real part of the pole position on the complex plane and the width corresponds to twice its imaginary part. In this case, the box diagrams corresponding to decays into two pseudoscalars

Table 1. The properties, (mass, width) [in units of MeV], of the 11 dynamically generated states and, if existing, of those of their PDG counterparts [10]. The association of the dynamically generated states with their experimental counterparts is determined by matching their mass, width, and decay pattern.

$I^G(J^{PC})$	Theory			PDG data		
	Pole position	Real axis		Name	Mass	Width
		$\Lambda_b = 1.4$ GeV	$\Lambda_b = 1.5$ GeV			
$0^+(0^{++})$	(1512,51)	(1523,257)	(1517,396)	$f_0(1370)$	1200~1500	200~500
$0^+(0^{++})$	(1726,28)	(1721,133)	(1717,151)	$f_0(1710)$	1724 ± 7	137 ± 8
$0^-(1^{+-})$	(1802,78)	(1802,49)		h_1		
$0^+(2^{++})$	(1275,2)	(1276,97)	(1275,111)	$f_2(1270)$	1275.1 ± 1.2	$185.0^{+2.9}_{-2.4}$
$0^+(2^{++})$	(1525,6)	(1525,45)	(1525,51)	$f_2'(1525)$	1525 ± 5	73^{+6}_{-5}
$1^-(0^{++})$	(1780,133)	(1777,148)	(1777,172)	a_0		
$1^+(1^{+-})$	(1679,235)	(1703,188)		b_1		
$1^-(2^{++})$	(1569,32)	(1567,47)	(1566,51)	$a_2(1320)$??		
$1/2(0^+)$	(1643,47)	(1639,139)	(1637,162)	K_0^*		
$1/2(1^+)$	(1737,165)	(1743,126)		$K_1(1650)$?		
$1/2(2^+)$	(1431,1)	(1431,56)	(1431,63)	$K_2^*(1430)$	1429 ± 1.4	104 ± 4

are not included. In the "real axis" method, the resonance parameters are obtained from the modulus squared of the amplitudes of the dominant channel of each state on the real axis (see Tables I, II, and III of Ref. [3]), where the mass corresponds to the energy at which the modulus squared has a maximum and the width corresponds to the difference between the two energies where the modulus squared is half of the maximum value. In this latter case, the box amplitudes are included. The results shown in Table I have been obtained using two different values of the box cut off, which serve to quantify the uncertainties related to this parameter.

Our treatment of the box amplitudes also enables us to obtain the decay branching ratios of the generated states into two pseudoscalar mesons [3]. The branching ratios are given in Tables 2, 3, and 4, in comparison with available data [10].

From Table 2, one can see that our results for the two f_2 states agree very well with the data. For the $f_0(1370)$, according to the PDG [10], the $\rho\rho$ mode is dominant. In our approach, however, the $\pi\pi$ mode is dominant, which is consistent with the results of Ref. [11] and the analysis of D. V. Bugg [12]. For the $f_0(1710)$, using the branching ratios given in Table 2, we obtain $\Gamma(\pi\pi)/\Gamma(K\bar{K}) < 1\%$ and $\Gamma(\eta\eta)/\Gamma(K\bar{K}) \sim 49\%$. On the other hand, the PDG gives the following averages: $\Gamma(\pi\pi)/\Gamma(K\bar{K}) = 0.41^{+0.11}_{-0.17}$, and $\Gamma(\eta\eta)/\Gamma(K\bar{K}) = 0.48 \pm 0.15$ [10]. Our calculated branching ratio for the $\eta\eta$ channel is in agreement with their average, while the ratio for the $\pi\pi$ channel is much smaller. However, we notice that the above PDG $\Gamma(\pi\pi)/\Gamma(K\bar{K})$ ratio is taken from the BES data on $J/\psi \to \gamma\pi^+\pi^-$ [13], which comes from a partial wave analysis that includes seven resonances. On the other hand, the BES data on $J/\psi \to \omega K^+K^-$ [14] give an upper limit $\Gamma(\pi\pi)/\Gamma(K\bar{K}) < 11\%$ at the 95% confidence level. Clearly more analysis is advised to settle the issue.

Compared to the decay branching ratios into two pseudoscalar mesons of the isospin 0 states, those of the isospin 1 states with spin either 0 or 2 are relatively small, as shown in Table 3.

Table 2. Branching ratios of the $f_0(1710)$, $f_0(1370)$, $f_2(1270)$, and $f_2'(1525)$ in comparison with data [10].

	$\Gamma(\pi\pi)/\Gamma(\text{total})$		$\Gamma(\eta\eta)/\Gamma(\text{total})$		$\Gamma(K\bar{K})/\Gamma(\text{total})$		$\Gamma(\text{VV})/\Gamma(\text{total})$	
	Our model	Data	Our model	Data	Our model	Data	Our model	Data
$f_0(1370)$	$\sim 72\%$		$< 1\%$		$\sim 10\%$		$\sim 18\%$	
$f_0(1710)$	$< 1\%$		$\sim 27\%$		$\sim 55\%$		$\sim 18\%$	
$f_2(1270)$	$\sim 88\%$	84.8%	$< 1\%$	$< 1\%$	$\sim 10\%$	4.6%	$< 1\%$	
$f_2'(1525)$	$< 1\%$	0.8%	$\sim 21\%$	10.4%	$\sim 66\%$	88.7%	$\sim 13\%$	

Table 3. Branching ratios of the a_0 and a_2 states.

	$\Gamma(K\bar{K})/\Gamma(\text{total})$	$\Gamma(\pi\eta)/\Gamma(\text{total})$	$\Gamma(\text{VV})/\Gamma(\text{total})$
a_0	$\sim 27\%$	$\sim 23\%$	$\sim 50\%$
a_2	$\sim 21\%$	$\sim 17\%$	$\sim 62\%$

Table 4. Branching ratios of the K_0^* and $K_2^*(1430)$ states in comparison with data [10].

	$\Gamma(K\pi)/\Gamma(\text{total})$		$\Gamma(K\eta)/\Gamma(\text{total})$		$\Gamma(\text{VV})/\Gamma(\text{total})$	
	Our model	Data	Our model	Data	Our model	Data
K_0^*	$\sim 65\%$		$\sim 9\%$		$\sim 26\%$	
$K_2^*(1430)$	$\sim 93\%$	49.9%	$\sim 5\%$	$< 1\%$	$\sim 2\%$	

In Table 4, one can see that the dominant decay mode of the $K_2^*(1430)$ is $K\pi$ both theoretically and experimentally. However, other modes, such as ρK, $K^*\pi$, and $K^*\pi\pi$, account for half of its decay width according to the PDG [10]. This is consistent with the fact that our $K_2^*(1430)$ is narrower than its experimental counterpart, as can be seen from Table 1.

The three spin 1 states with the quantum numbers of h_1, b_1 and K_1, do not decay into two pseudoscalars in our approach since the box diagrams do not contribute as explained in Ref. [3].

It is interesting to note that out of the 21 combinations of strangeness, isospin and spin, we have found resonances only in nine of them. In all the "exotic" channels, from the point of view that they cannot be formed from $q\bar{q}$ combinations, we do not find dynamically generated resonances, including the three (strangeness=0, isospin=2) channels, the three (strangeness=1, isospin=3/2) channels, the six strangeness=2 channels (with either isospin=0 or isospin=1). On the other hand, there do exist some structures on the real axis. For instance, in the (strangeness=0, isospin=2) channel, one finds a dip around $\sqrt{s} = 1300$ MeV in the spin=0 channel, and a broad bump in the spin=2 channel around $\sqrt{s} = 1400$ MeV, as can be clearly seen from Fig. 7 of Ref. [3]. In the (strangeness=1, isospin=3/2) and (strangeness=2, isospin=1) channels, one observes similar structures occurring at shifted energies due to the different masses of the ρ and the K^*, as can be seen from Figs. 8 and 9 of Ref. [3]. However, these states do not correspond to poles on the complex plane, and hence, according to the common criteria, they do not qualify as resonances.

4. Further applications

One of the appealing features of the coupled channel unitary approach is that one can make many predictions concerning properties of the dynamically generated resonances, as well as rates of production in different reactions. This is so because these states are molecular states of ground state meson or baryon components, with well known couplings to external sources which allow one to obtain the coupling of the resonances to these sources. As examples there has been work on the radiative decay of the resonances reported here to two photons, or one vector and one photon in [15, 16], with good agreement with experiment whenever comparison is possible. Similarly, in [17] there has been work on the decay modes of the J/ψ into a vector meson (ϕ, ω or K^{*0}) and a tensor meson, taking into account the nature of the $f_2(1270)$, $f_2'(1525)$, $\bar{K}_2^{*0}(1430)$ as dynamically generated states as has been discussed here. The agreement of the predictions, free of any parameter, with the data is fair. Also, the radiative decay modes of the J/ψ into a photon and one of the tensor mesons $f_2(1270)$, $f_2'(1525)$, as well as the scalar ones $f_0(1370)$ and $f_0(1710)$ has been investigated in [18] and the agreement with data is also acceptable. Similarly, good agreemnet with the scarce data is found in [19] for the decay of these resonances into a photon and a pseudoscalar meson. In this work it is also found that the association of the a_2 state found to the $a_2(1320)$ is a fair assumption.

5. Summary and conclusions

We have presented the results of a study of vector meson-vector meson interaction using a unitary approach. Employing the coupled-channel Bethe-Salpeter equation to unitarize the transition amplitudes provided by the hidden-gauge Lagrangians, we find that 11 states get dynamically generated in nine strangeness-isospin-spin channels. Among them, five states are associated to those reported in the PDG, i.e., the $f_0(1370)$, the $f_0(1710)$, the $f_2(1270)$, the $f_2'(1525)$, the $K_2^*(1430)$. The association of two other states, the $a_2(1320)$ and the $K_1(1650)$, are likely, but less certain. Four of the 11 dynamically generated states can not be associated with any known states in the PDG. Finally, we also reported on very recent developments, where application of the results to different decays processes has been made. The consistent agreement with data of the results obtained gives progressively strong support for the nature of these resonances as being dynamically generated from the vector-vector interaction.

6. References

[1] J. A. Oller, E. Oset and A. Ramos, Prog. Part. Nucl. Phys. **45**, 157 (2000).
[2] R. Molina, D. Nicmorus and E. Oset, Phys. Rev. D **78**, 114018 (2008).
[3] L. S. Geng and E. Oset, Phys. Rev. D **79**, 074009 (2009).
[4] R. Molina, H. Nagahiro, A. Hosaka, E. Oset, Phys. Rev. **D80**, 014025 (2009).
[5] P. Gonzalez, E. Oset and J. Vijande, Phys. Rev. C **79**, 025209 (2009).
[6] S. Sarkar, B. -X. Sun, E. Oset, M. J. Vicente Vacas, Eur. Phys. J. **A44**, 431-443 (2010).
[7] E. Oset, A. Ramos, Eur. Phys. J. **A44**, 445-454 (2010).
[8] M. Bando, T. Kugo, S. Uehara, K. Yamawaki and T. Yanagida, Phys. Rev. Lett. **54**, 1215 (1985); M. Bando, T. Kugo and K. Yamawaki, Phys. Rept. **164**, 217 (1988); U. G. Meissner, Phys. Rept. **161**, 213 (1988).
[9] A. I. Titov, B. Kampfer and B. L. Reznik, Eur. Phys. J. A **7**, 543 (2000); A. I. Titov, B. Kampfer and B. L. Reznik, Phys. Rev. C **65**, 065202 (2002).
[10] C. Amsler *et al.* [Particle Data Group], Phys. Lett. B **667**, 1 (2008).
[11] M. Albaladejo and J. A. Oller, Phys. Rev. Lett. **101**, 252002 (2008).
[12] D. V. Bugg, Eur. Phys. J. C **52**, 55 (2007).
[13] M. Ablikim *et al.*, Phys. Lett. B **642**, 441 (2006).
[14] M. Ablikim *et al.* [BES Collaboration], Phys. Lett. B **603** (2004) 138.
[15] H. Nagahiro, J. Yamagata-Sekihara, E. Oset, S. Hirenzaki and R. Molina, Phys. Rev. D **79**, 114023 (2009)
[16] T. Branz, L. S. Geng, E. Oset, Phys. Rev. **D81**, 054037 (2010).
[17] A. Martinez Torres, L. S. Geng, L. R. Dai, B. X. Sun, E. Oset and B. S. Zou, Phys. Lett. B **680**, 310 (2009)
[18] L. S. Geng, F. K. Guo, C. Hanhart, R. Molina, E. Oset, B. S. Zou, Eur. Phys. J. **A44**, 305-311 (2010).
[19] R. Molina, H. Nagahiro, A. Hosaka, E. Oset, Phys. Rev. **D83**, 094030 (2011).

Predictions for NDK, $\bar{K}DN$ and $ND\bar{D}$ molecules

M Bayar [1, 2], C W Xiao [1] and E Oset [1]

[1] *Departamento de Física Teórica and IFIC, Centro Mixto Universidad de Valencia-CSIC, Institutos de Investigación de Paterna, Aptdo. 22085, 46071 Valencia, Spain*
[2] *Department of Physics, Kocaeli University, 41380 Izmit, Turkey*
(Dated: January 26, 2012)

In this work baryon systems made of three hadrons which contain one nucleon and one D meson, and in addition another meson, $\bar{D}, K$ or $\bar{K}$, are investigated using the Fixed Center Approximation to the Faddeev equations. In this work we use $\Lambda_c(2595)$, X(3700) and $D_{s0}^*(2317)$ bound states as a cluster and a third particle scattering form that clusters. In all cases we find bound states and quasibound states.

PACS numbers: 14.40.Lb; 14.20.Lq; 21.45.-v

I. INTRODUCTION

The charm sector has not yet been explored for such three body systems and this is the first incursion in that field. For this purpose we have selected systems that have a nucleon and a D meson. The DN system, in collaboration with coupled channels, leads to the formation of a dynamically generated state, the $\Lambda_c(2595)$ [1–4]. On top of it we add a $K, \bar{K}$ or $\bar{D}$ meson and we study the stability of the system. The case of scattering of N on the DK cluster, which is known to generate the $D_{s0}^*(2317)$ [5–7], is also considered. On the other hand, the $D\bar{D}$ system leads to a bound state in isospin I=0 [7], which might have already been observed [8]. We add a nucleon to it and study the interaction of the three body system.

II. MULTI-BODY INTERACTION FORMALISM

The FCA to the Faddeev equations are an effective tool to investigate multi-hadron interaction [9–12]. They are particularly suited to study system where a pair of particles cluster together and the cluster is not much modified by the third particle.

The FCA approximation to Faddeev equations assumes a pair of particles (1 and 2) forming a cluster. Then particle 3 interacts with the components of the cluster, undergoing all possible multiple scattering with those components. This is shown in Fig. 1. First, one defines two partition functions T_1, T_2 which sum all diagrams of the series of Fig. 1 which begin with the interaction of particle 3 with particle 1 of the cluster (T_1), or with the particle 2 (T_2), then the total scattering

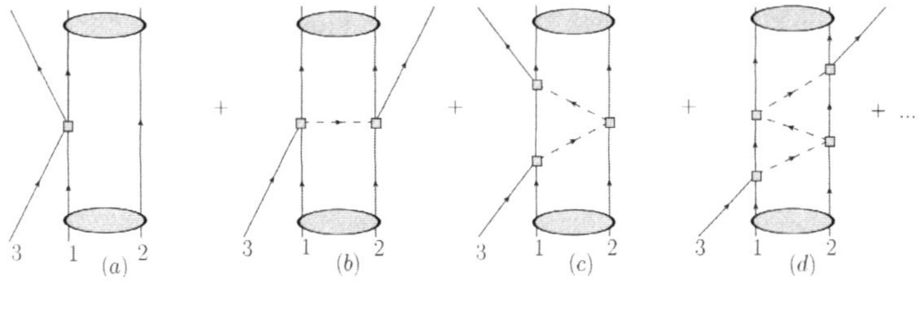

FIG. 1: Diagrammatic representation of the FCA to Faddeev equations. The equivalent diagrams where the particle 3 interacts first with the particle 2 should be added.

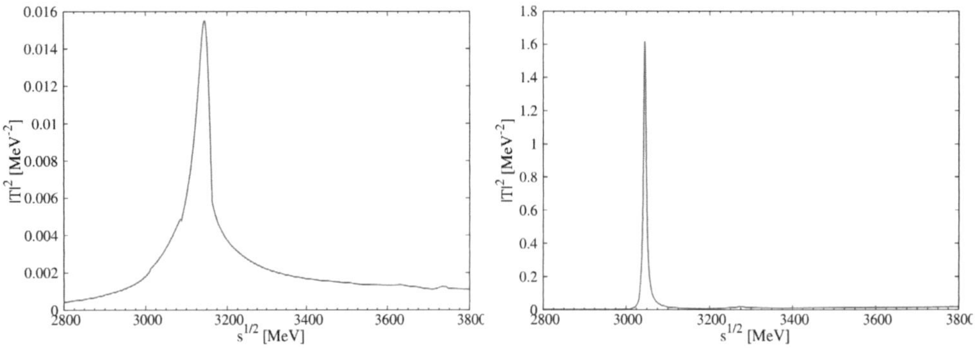

FIG. 2: Modulus squared of the scattering amplitude for $\bar{K}\Lambda_c(2595)$ (left) and $ND^*_{s0}(2317)$ (right).

amplitude is sum of T_1 and T_2.

$$T_1 = t_1 + t_1 G_0 T_2 \; ; \quad T_2 = t_2 + t_2 G_0 T_1 \; ; \quad T = T_1 + T_2, \tag{1}$$

where T is the total three-body scattering amplitude that we are looking for. The amplitudes t_1 and t_2 represent the unitary scattering amplitudes with coupled channels for the interactions of particle 3 with particle 1 and 2, respectively. Besides, G_0 is the propagator of particle 3 between the components of the two-body system. For the detailed formalism one can look at the original paper [13].

III. THE FORMALISM FOR THE FCA IN THE THREE BODY SYSTEM

Our strategy proceeds as follows: first we generate the resonance or bound state in the compound system, then calculate the form factor and G_0 propagator and take the t_1 and t_2 amplitudes from the unitary coupled channel approach, finally the total scattering amplitude T is evaluated.

We investigated two three body scattering, $N-(DK)_{D^*_{s0}(2317)}$ and $K-(DN)_{\Lambda_c(2595)}$, in the NDK system since the resonance $D^*_{s0}(2317)$ is dynamically generated in $I=0$ from KD scattering and the $\Lambda_c(2595)$ is produced in $I=0$ from the DN interaction. The result of the $ND^*_{s0}(2317)$ system is represented in Fig. 2. We found a peak around 3050 MeV which is about 200 MeV below the $D^*_{s0}(2317)$ and N threshold. This reflects the strong attraction in the DN system that leads to the $\Lambda_c(2595)$. The width of the state is smaller than 10 MeV. We do not find a counterpart in the PDG and the quantum numbers, with positive strangeness, correspond to an exotic state. But the combination of $K-(DN)_{\Lambda_c(2595)}$ is too small comparing with the $ND^*_{s0}(2317)$ configuration. This means that the $K(DN)$ configuration in the wave function of the KDN system has a very small weight. Hence, we predict a bound state of NDK mostly made of a N orbiting around a bound DK cluster forming the $D^*_{s0}(2317)$. In Fig. 2 we also show the results of $|T|^2$ for the $\bar{K}\Lambda_c(2595)$ scattering. We find a peak around 3150 MeV, slightly above the threshold of the $\Lambda_c(2595)+\bar{K}$ mass (3088 MeV) and below the threshold of the $\bar{K}DN$ system (3298 MeV). The width of the peak is about 50 MeV. Finally, we obtain the T matrix for the $ND\bar{D}$ interaction by means Eq. (1), and show the results of $|T|^2$ in Fig. 3. From this figure we can see that there is a clear peak of $|T|^2$ around 4400 MeV and the width is very small, less than 10 MeV.

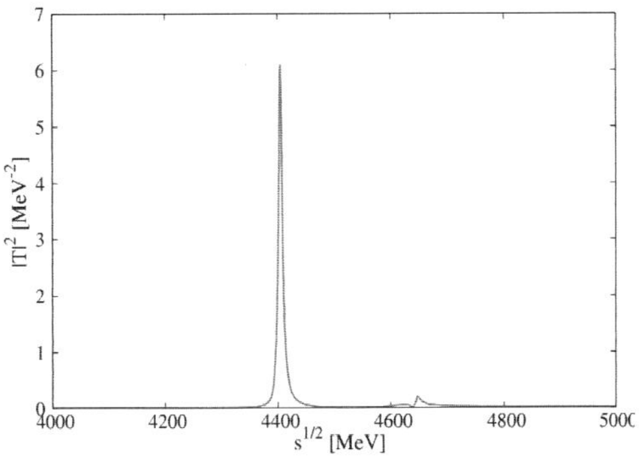

FIG. 3: Modulus squared of the $NX(3700)$ scattering amplitude.

IV. CONCLUSIONS

We have investigated the $\bar{K}DN$, NDK and $ND\bar{D}$ three body systems. In all cases we find bound or quasibound states, relatively narrow, with energies 3150 MeV, 3050 MeV and 4400 MeV, respectively. All these states have $J^P = 1/2^+$ and isospin $I = 1/2$ and differ by their charm or strangeness content, $S = -1, C = 1$, $S = 1, C = 1$, $S = 0, C = 0$, respectively. The first state could perhaps be associated to the $\Xi(3123)$, which has unknown J^P, but the width obtained is a bit too large. The second state is of exotic nature and there is no counterpart in the PDG. The third state is a regular N^* state as to quantum numbers, but it contains hidden charm. It lies in an energy region where baryon states have not yet been investigated. We are thus making predictions in the frontier of this field and hope that with the coming Facilities of FAIR, or the BELLE upgrade, such states can be systematically studied and that our predictions can be confirmed.

Acknowledgments

This work is partly supported by DGICYT contract FIS2006-03438, the Generalitat Valenciana in the program Prometeo and the EU Integrated Infrastructure Initiative Hadron Physics Project under Grant Agreement n.227431. M. Bayar acknowledges support through the Scientific and Technical Research Council (TUBITAK) BIDEP-2219 grant.

[1] J. Hofmann and M. Lutz, Nucl.Phys. **A763**, 90 (2005).
[2] T. Mizutani and A. Ramos, Phys.Rev. **C74**, 065201 (2006).
[3] L. Tolos, A. Ramos, and T. Mizutani, Phys.Rev. **C77**, 015207 (2008).
[4] C. Garcia-Recio, V. Magas, T. Mizutani, J. Nieves, A. Ramos, et al., Phys.Rev. **D79**, 054004 (2009).
[5] J. Hofmann and M. Lutz, Nucl.Phys. **A733**, 142 (2004).
[6] F.-K. Guo, P.-N. Shen, H.-C. Chiang, R.-G. Ping, and B.-S. Zou, Phys.Lett. **B641**, 278 (2006).
[7] D. Gamermann, E. Oset, D. Strottman, and M. Vicente Vacas, Phys.Rev. **D76**, 074016 (2007).
[8] D. Gamermann and E. Oset, Eur.Phys.J. **A36**, 189 (2008).
[9] J.-J. Xie, A. Martinez Torres, and E. Oset, Phys.Rev. **C83**, 065207 (2011).
[10] M. Bayar, J. Yamagata-Sekihara, and E. Oset, Phys.Rev. **C84**, 015209 (2011).
[11] L. Roca and E. Oset, Phys.Rev. **D82**, 054013 (2010).
[12] J.-J. Xie, A. Martinez Torres, E. Oset, and P. Gonzalez, Phys.Rev. **C83**, 055204 (2011).
[13] C. Xiao, M. Bayar, and E. Oset, Phys.Rev. **D84**, 034037 (2011).

Journal of Physics: Conference Series **348** (2012) 012006 doi:10.1088/1742-6596/348/1/012006

Application of the Maximum Entropy Method to QCD sum rules[1]

Makoto Oka and Philipp Gubler

Department of Physics, Tokyo Institute of Technology, Meguro, Tokyo 152-8551, Japan

E-mail: oka@th.phys.titech.ac.jp

Abstract. A new method of analyzing QCD sum rules employing the Maximum Entropy Method, is introduced and is applied successfully to the rho meson, the nucleon and the charmonium at finite temperature. This method enables us to directly obtain the spectral function from the sum rules, without having to introduce any model assumption about its functional form. In the nucleon sum rule, we show that the Gaussian sum rule successfully reproduces the ground state. Dependences on the interpolating field operator are discussed. Finite temperature effects for the charmonium sum rule are incorporated by changes of various gluonic condensates, extracted from lattice QCD. As a result, we find that both J/ψ and η_c dissolve into the continuum already at temperatures around $1.0 \sim 1.2\,T_c$.

1. Introduction

Recent discoveries of many new (exotic) hadron states have revitalized experimental and theoretical studies of the hadron spectroscopy. We have realized that the excited hadrons are much stranger than expected and require new concepts, such as multi-quarks, molecular resonances, dynamically generated resonances, chiral partners, and so on. It is quite interesting to establish a universal view of both the ground-state and excited hadrons of all variations in terms of the fundamental degrees of freedom in QCD. Such a universal picture will incorporate the dynamics of color-singlet hadrons as well as the mechanism of quark-gluon confinement.

The hadron spectrum is also useful to investigate the symmetry and dynamical properties of the QCD vacuum. Possible changes of hadron properties at finite temperature and/or density will provide us with crucial information on how and when chiral symmetry is restored, how quarks and gluons are released from hadron, and what the properties of the matter after phase transition in QCD are.

For these goals, the spectral functions of hadrons play key roles and should be computed directly from QCD. Lattice QCD is obviously a strong non-perturbative method, while the QCD sum rule is also a powerful tool as a complimentary approach. In the hadron spectroscopy, the QCD sum rule is suitable for $m_q \sim 0$ and the continuum limit, while lattice QCD is appropriate for $L^{-1} < m_q < a^{-1}$, where L is the lattice size and a is the lattice spacing.

The QCD sum rule is a method to derive relations between the QCD vacuum condensates, such as $\langle \bar{q}q \rangle$, $\langle G^2 \rangle$, $\langle \bar{q}\sigma \cdot Gq \rangle$ and so on, and the hadron spectral function and observables, such as mass, coupling, and so on. The QCD sum rule works very well to derive the masses of stable hadrons, and sharp resonance states, *ex.* $\rho(770)$, $\Delta(1232)$, etc. It has been applied to the coupling constants of hadrons, $g_{\pi NN}$, g_A etc.

[1] A part of this work has been done in collaboration with Kenji Morita and Keisuke Ohtani.

In this report, we discuss a new method in studying the spectral function using QCD sum rules. We recently proposed[1] to apply the Maximum Entropy Method (MEM) to QCD sum rules and estimate the spectral function directly without assuming its functional form. This method was developed first for the rho meson sum rule in order to check the method[1], and has been successfully applied to the nucleon and its excited states[2, 3], and also to the finite temperature behavior of quarkonia[4, 5].

2. Maximum Entropy Method in QCD Sum Rules

In this section, the main ideas of the MEM procedure are briefly reviewed. (For the details, see [1, 6, 7].) The QCD sum rule, in the standard form after the Borel transform, is given by an integral of the spectral function, $\rho(\omega)$, as

$$G_{\rm OPE}(M) = \int_0^\infty d\omega K(M,\omega)\rho(\omega), \tag{1}$$

where $K(M,\omega)$ is the kernel given, in the case of the Borel sum rule, by $K(M,\omega) = 2\omega e^{-\omega^2/M^2}$. $G_{\rm OPE}(M)$ is the correlator calculated by the operator product expansion (OPE) at the deep Euclidean region, and is known only with limited accuracy and in a limited range of the Borel mass M. Then, the problem of obtaining $\rho(\omega)$ from $G_{\rm OPE}(M)$ is ill-posed and Eq. (1) cannot be analytically inverted.

The MEM approach now uses Bayes' theorem, by which additional information about $\rho(\omega)$ such as positivity and/or its asymptotic behavior at small or large energies can be included into the analysis in a systematic way and by which one can finally deduce the most probable from of $\rho(\omega)$. Bayes' theorem can be expressed as

$$P[\rho|GH] = \frac{P[G|\rho H]P[\rho|H]}{P[G|H]}, \tag{2}$$

where the prior knowledge about $\rho(\omega)$ is denoted as H and $P[\rho|GH]$ stands for the conditional probability of $\rho(\omega)$ given $G_{\rm OPE}(M)$ and H. Dropping the constant term $P[G|H]$ in the denominator because it does not depend on $\rho(\omega)$ and maximizing the remaining functional will give the most probable $\rho(\omega)$. $P[G|\rho H]$ is the "likelihood function" and can be constructed as

$$P[G|\rho H] = e^{-L[\rho]}, \qquad L[\rho] = \frac{1}{2(M_{\rm max}-M_{\rm min})}\int_{M_{\rm min}}^{M_{\rm max}} dM \frac{[G_{\rm OPE}(M) - G_\rho(M)]^2}{\sigma^2(M)}, \tag{3}$$

where $G_\rho(M)$ is defined as on the right hand side of Eq. (1), and therefore implicitly depends on $\rho(\omega)$. The function $\sigma(M)$ describes the uncertainty of $G_{\rm OPE}(M)$ at the corresponding value of the Borel mass.

On the other hand, $P[\rho|H]$ is called the "prior probability" and is given as

$$P[\rho|H] = e^{\alpha S[\rho]}, \qquad S[\rho] = \int_0^\infty d\omega\Big[\rho(\omega) - m(\omega) - \rho(\omega)\log\Big(\frac{\rho(\omega)}{m(\omega)}\Big)\Big], \tag{4}$$

where $S[\rho]$ is called the Shannon-Jaynes entropy. The function $m(\omega)$ is called "default model", which will be the result for $\rho(\omega)$ by MEM when no data $G_{\rm OPE}(M)$ is available. The default model can be used to incorporate known information about $\rho(\omega)$ into the calculation.

Assembling the functions given above, we get the final form for the probability $P[\rho|GH]$:

$$P[\rho|GH] \propto P[G|\rho H]P[\rho|H] = e^{Q[\rho]}, \qquad Q[\rho] \equiv \alpha S[\rho] - L[\rho]. \tag{5}$$

The form of $\rho(\omega)$ that maximizes $Q[\rho]$ is obtained numerically by the use of the Bryan algorithm [8], which is widely used in MEM studies.

Once we have found $\rho_\alpha(\omega)$ which maximizes $Q[\rho]$ for a fixed value of α, this parameter is integrated out by averaging $\rho(\omega)$ over a range of values of α. The explicit procedure for this step is explained in [1]. The result this last integration then leads to our final result $\rho_{\rm out}(\omega)$.

3. Some Results from the MEM QCD Sum Rules

In the following sections, we give the highlights from the QCD sum rule analyses using the Maximum Entropy Method. The new MEM analysis was first tested in the rho-meson sum rule[1]. The obtained spectral function shows a clear ρ meson peak, which indicates that the MEM is useful in the sum rule analysis. We were able to extract the ρ meson mass, which lies within about 10% of the experimental value, and also the coupling strength within about 30%. On the other hand, it turns out that the OPE correlator, with the current ambiguity of the condensates and the higher order terms, is not accurate enough to extract the width of the ρ meson.

3.1. Nucleon Sum Rule

The MEM analysis has been applied to the nucleon sum rule[2]. We have investigated two kinds of sum rules, namely the frequently used Borel sum rule and the less known Gaussian sum rule[9, 10]. In fact, in the case of the Borel sum rule, we find that due to the properties of the kernel and the slow OPE convergence, it is difficult to extract much meaningful information on the nucleon ground state. As an alternative, we have formulated the Gaussian sum rule and found that it allows us to extract more information on the spectral function and enables us to reconstruct the nucleon ground state with reasonable precision.

The derivation of the Gaussian sum rule is given in [2]. It ends up with a two-parameter integral of the spectral function,

$$G_\rho(s,\tau) = \int_0^\infty d\omega K(s,\tau,\omega)\rho(\omega), \qquad K(s,\tau,\omega) = \frac{2\omega^5}{\sqrt{4\pi\tau}} e^{-\frac{(\omega^2-s)^2}{4\tau}}. \tag{6}$$

where the spectral function ρ is normalized by its high energy behavior, namely by a factor $1/\omega^5$. The new parameters, s and τ, play the role of the Borel mass, and are to be taken so that the OPE convergence and the pole dominance are satisfied. The allowed region of the parameter values happens to be larger because we have two parameters.

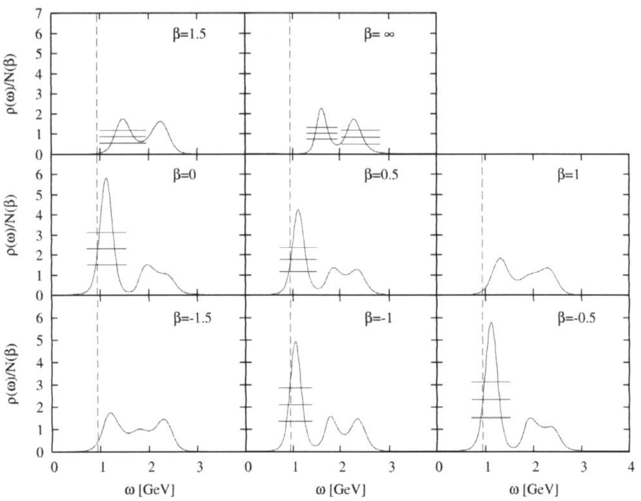

Figure 1. The nucleon spectral functions for various values of β. They are normalized by $N(\beta) \equiv (5 + 2\beta + 5\beta^2)/(128(2\pi)^4)$. The horizontal lines indicate the error estimated from the MEM error analysis. The vertical dashed lines show the nucleon mass position.

For the nucleon, there are two independent local interpolating field operators,

$$\eta_1(x) = \epsilon^{abc}(u^{Ta}(x)C\gamma_5 d^b(x))u^c(x), \quad \text{and} \quad \eta_2(x) = \epsilon^{abc}(u^{Ta}(x)Cd^b(x))\gamma_5 u^c(x). \tag{7}$$

Here, abc are color indices and C is the charge conjugation matrix. Then the general interpolating operator is given by $\eta(x) = \eta_1(x) + \beta\eta_2(x)$ with a parameter β. The $\beta = -1$ operator is the so-called "Ioffe current", which is often used in the nucleon sum rule.

The main results of the MEM analyses of the Gaussian sum rule are shown in Fig.1. One clearly sees the ground state nucleon as a peak for $-1 \le \beta \le 0.5$. We have also found that the strength and the shape of the nucleon peak is nearly independent of the parameter β and that the nucleon peak vanishes for the pure η_2 operator. Thus we conclude that, in agreement with findings of lattice QCD[11], the nucleon ground state couples only to the interpolating field η_1, but not (or only very weakly) to η_2. Another peak structure is seen around 1.6 GeV in the spectral function corresponding to η_2, which suggests that some nucleon resonance in this region couples to η_2. In order to clarify the nature of this second peak, we have performed the analysis of the parity projected sum rule[3]. We find that the second peak has the negative parity so that it might correspond to the N*(1535) and/or (1650) resonances.

3.2. Charmonium Spectrum at Finite Temperature

The behavior of the charmonium spectrum in the pseudoscalar and vector channels is studied by the MEM[4]. The results of the OPE (For details, see [12]) include the dimension-4 scalar gluon condensate, $G_0 = \langle(\alpha_s/\pi)G^a_{\mu\nu}G^{a\mu\nu}\rangle_T$, twist-2 condensate defined by $(u^\mu u^\nu - \frac{1}{4}g^{\mu\nu})G_2 = \langle(\alpha_s/\pi)G^{a\mu\sigma}G^{a\nu}_{\sigma}\rangle_T$ (u^μ being the four velocity of the heat bath), and the dimension-6 condensate, $G_3 = \langle g^3 f^{abc}G^{a\nu}_{\mu}G^{b\lambda}_{\nu}G^{c\mu}_{\lambda}\rangle_T$.

We assume that all finite temperature effects appear only in the gluon condensates in the OPE, which were expressed in terms of the energy density and pressure, $\epsilon(T)$ and $p(T)$[13], as

$$G_0 = G_0^{\text{vac}} - \frac{8}{11}[\epsilon(T) - 3p(T)], \qquad G_2 = -\frac{\alpha_s(T)}{\pi}[\epsilon(T) + p(T)]. \tag{8}$$

The functions $\epsilon(T)$, $p(T)$ and $\alpha_s(T)$ were then extracted from quenched lattice QCD data, for which we use the values given in [12]. It should be noted that due to the sudden change of $\epsilon(T)$ and $p(T)$ above the critical temperature, G_0 and G_2 show an abrupt decrease around T_c. The temperature dependence of the dimension 6 condensate is evaluated by the use of a relation based on the dilute instanton gas model. Its effect on the behavior of the charmonium states, however, turns out to be small.

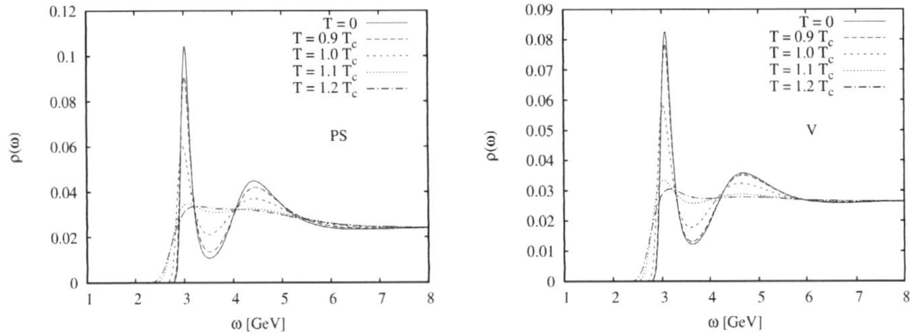

Figure 2. The spectral functions in the pseudoscalar (left) and vector (right) channels at temperatures between $T = 0.9\,T_c$ and $T = 1.2\,T_c$.

The Borel mass region is chosen so that the magnitude of the dimension 6 term is smaller than 20% of the whole OPE expression, and the upper limit of the Borel mass is 3 GeV. For the other choices of the various parameters, see [4]. The results are shown in Fig. 2. At $T = 0$, the ground state peaks for both $\eta_c(J = 0)$ and $J/\psi(J = 1)$ are clearly extracted. The η_c (J/ψ) peak lies at 3.02 (3.06) GeV. We thus find that while the peak positions reproduce the experimental values with a precision of about 50 MeV, the hyperfine splitting is underestimated. These findings are in agreement with earlier QCD sum rule studies [14], showing that the MEM analysis works properly.

Fig. 2 also shows the spectral functions in hot media between $T = 0.9\,T_c$ and $T = 1.2\,T_c$. One observes a sudden change in the spectral function just above T_c, which is related to the abrupt increase of

$\epsilon(T)$ and $p(T)$ at this temperature. The detailed behavior of the spectral function can be summarized as follows: first, the peaks experience a shift of about 50 MeV to lower energies and then suddenly "melt" into the continuum. For η_c, the melting temperature is $T = 1.1\,T_c$, while for J/ψ it lies slightly higher, namely at $T = 1.2\,T_c$. The shift of the peak observed here is perfectly consistent with the previous QCD sum rule analysis which used a more conventional analysis method [13, 12].

We would like to mention here that, unlike in similar studies using lattice QCD, the number of data points that can be used for the MEM analysis are not reduced when the temperature is increased, allowing a reliable extraction of the spectral function at finite temperatures. Moreover, we have checked that the additional uncertainties that enter into the analysis at finite temperatures due to the functions $\epsilon(T)$ and $p(T)$ do not cause a substantial reduction of the peaks. Thus, we conclude that the disappearance of the peaks shown in Fig. 2 is a physical effect and not caused by an artifact of the MEM analysis.

The present result is in quantitative disagreement to the corresponding melting temperature observed in earlier lattice calculations of correlators at Euclidean time [15]. It, however, has to be remarked that our results are in fact consistent with the latest lattice results [16], finding the peaks of η_c and J/ψ to be largely distorted between 0.73 T_c and 1.46 T_c. It remains to be seen whether the difference of the melting temperature is of physical importance, or if it will disappear when both methods will become more accurate in the future.

A similar results are obtained in the analysis of the bottomonium spectra at finite temperature[5]. There, we find a higher disappearance temperature and also indication that the excited bottomonia melt away before the ground state does.

4. Conclusions

The Maximum Entropy Method applied to the analysis of QCD sum rules is found to be very powerful and useful in extracting the spectral functions of the ρ meson, the nucleon, and the charmonium at finite temperature. The advantage of the MEM method is that one need not assume the existence of the sharp peak (pole) in the spectral function beforehand. Thus, we can reliably investigate the disappearance of the charmonium peak at high temperature, and also the operator dependences of the nucleon sum rules. We believe that the MEM QCD-SR becomes a standard method for analyses of the hadron spectrum.

Acknowledgments

We thank Kenji Morita, Keisuke Ohtani and Kei Suzuki for fruitful collaborations regarding the subjects in this article. This work is supported by KAKENHI under Contract No. 22105503. P.G. gratefully acknowledges the support by the Japan Society for the Promotion of Science for Young Scientists (Contract No. 21.8079).

References

[1] P. Gubler and M. Oka, Prog. Theor. Phys. **124**, 995 (2010).
[2] K. Ohtani, P. Gubler and M. Oka, Eur. Phys. J. **A47** (2011) 114.
[3] K. Ohtani, P. Gubler and M. Oka, in preparation.
[4] P. Gubler, K. Morita and M. Oka, Phys. Rev. Lett. **107** (2011) 092003.
[5] K. Suzuki, P. Gubler, K. Morita and M. Oka, in preparation.
[6] M. Jarrel and J.E. Gubernatis, Phys. Rep. **269**, 133 (1996).
[7] M. Asakawa, T. Hatsuda, and Y. Nakahara, Prog. Part. Nucl. Phys. **46**, 459 (2001).
[8] R.K. Bryan, Eur. Biophys. J. **18**, 165 (1990).
[9] R.A. Bertlmann, G. Launer and E. de Rafael, Nucl. Phys. **B250**, 61 (1985).
[10] G. Orlandini, T.G. Steele and D. Harnett, Nucl. Phys. **A686**, 261 (2001).
[11] D.B. Leinweber, Phys, Rev. D **51**, 6383 (1995); S. Sasaki, T. Blum and S. Ohta, Phys. Rev. D **65**, 074503 (2002).
[12] K. Morita and S.H. Lee, Phys. Rev. D **82**, 054008 (2010).
[13] K. Morita and S.H. Lee, Phys. Rev. Lett. **100**, 022301 (2008).
[14] J. Marrow, J. Parker and G. Shaw, Z. Phys. C **37**, 103 (1987).
[15] M. Asakawa and T. Hatsuda, Phys. Rev. Lett. **92**, 012001 (2004).
[16] H.-T. Ding *et al.*, PoS **LAT2010**, 180 (2010).

Journal of Physics: Conference Series **348** (2012) 012007 doi:10.1088/1742-6596/348/1/012007

Exotic Charmonium and Bottomonium-like Resonances

Fernando S. Navarra[1], Marina Nielsen[1] and Jean-Marc Richard[2]
[1]Instituto de Física, Universidade de São Paulo,
Caixa Postal 66318, 05389-970 São Paulo, SP, Brazil
[2]Université de Lyon and Institut de Physique Nucléaire de Lyon, IN2P3-CNRS-UCBL
4, rue Enrico Fermi, F-69622 Villeurbanne, France

E-mail: navarra@if.usp.br, mnielsen@if.usp.br, j-m.richard@ipnl.in2p3.fr

Abstract. Many new states in the charmonium and bottomonium mass region were recently discovered by the BaBar, Belle and CDF Collaborations. We use the QCD Sum Rule approach to study the possible structure of some of these states. In particular we identify the recently observed bottomonium-like resonance $Z_b^+(10610)$ with the first excitation of the tetraquark $X_b(1^{++})$, the analogue of the $X(3872)$ state in the charm sector.

1. Introduction

Most of the new charmonium states discovered in recent years at the B factories and at the Tevatron, called X, Y, Z particles, do not seem to have a simple $c\bar{c}$ structure. Their production mechanism, masses, decay widths, spin-parity assignments and decay modes have been discussed in some reviews [1, 2, 3]. Although the masses of these states are above the corresponding thresholds of decays into a pair of open charm mesons, they decay into J/ψ or ψ' plus pions, which is unusual for $c\bar{c}$ states. Besides, their masses and decay modes are not in agreement with the predictions of potential models, which, in general, describe very well $c\bar{c}$ states. For these reasons, they are considered as candidates for exotic states such as hybrid, molecular or tetraquark states, with a more complex structure than the simple quark-antiquark states.

In the bottomonium mass region, the first particle that does not seem to have a simple $b\bar{b}$ structure was the $Y_b(10890)$, observed by the Belle Collaboration [4]. The Belle Collaboration also reported the observation of two charged narrow structures in the $\pi^{\pm}\Upsilon(nS)$ $(n = 1, 2, 3)$ and $\pi^{\pm}h_b(mP)$ $(m = 1, 2)$ mass spectra of the $\Upsilon(5S) \to \Upsilon(nS)\pi^{\pm}$ and $\Upsilon(5S) \to h_b(mP)\pi^{\pm}$ decay processes [5]. These narrow structures were called $Z_b(10610)$ and $Z_b(10650)$. Analysis of angular distribution favors the quantum numbers $I^G(J^P) = 1^+(1^+)$ for both states. Since the Z_b are charged states, they are ideal candidates for exotic four-quark states, like the $Z^+(4430)$, also observed by the Belle Collaboration in $B^+ \to K\psi'\pi^+$ through its decay into $\psi'\pi^+$ [6].

2. Molecular States

As pointed out by the Belle Collaboration, the proximity of the $B\bar{B}^*$ and $B^*\bar{B}^*$ thresholds and the $Z_b(10610)$ and $Z_b(10650)$ masses suggests that these states could be interpreted as weakly bound $B\bar{B}^*$ and $B^*\bar{B}^*$ states [7, 8, 9, 10, 11]. They have also been interpreted as cusps at the $B\bar{B}^*$ and $B^*\bar{B}^*$ thresholds [12, 13] and as tetraquark states [14].

The molecular picture was also used to understand the well established $X(3872)$, which was considered to be a $D\bar{D}^*$ molecule. Of course, the actual physical state is probably rather complex, with a short-range component of $(c\bar{c})$ nature and a long-range component with a charmed meson and an anticharmed meson almost bound by attractive Yukawa forces. The idea of molecules with hidden charm has been proposed long ago by Okun and Voloshin [15]. A molecular interpretation was proposed for some high-lying 1^{--} charmonium resonances [16], due to puzzling branching ratios into $D\bar{D}$, $D\bar{D}^*$ and $D^*\bar{D}^*$, which turned out to be due to the node structure of these states as radial excitations of the J/ψ [17]. The possibility of hidden-charm meson molecules has been revisited in the 90s by Törnqvist [18], Ericson and Karl [19] and Manohar and Wise [20], and further developed by several authors after the discovery of the $X(3872)$ [2]. The main idea is that the Yukawa interaction, that successfully binds nuclei, is not restricted to the nucleon-nucleon interaction. The exchange of light mesons also generates a potential between flavored mesons, which is sometimes attractive. Although usually weaker than the proton-neutron interaction that binds the deuteron, it is probed by much heavier particles, and thus can lead to bound states with a binding energy of a few MeV, or even a few tens of MeV, as shown by Törnqvist [18]. For the molecular $B\bar{B}$ states, the study shows that the energy of isoscalars $B\bar{B}^*$ with $J^{PC} = 0^{-+}$, 1^{++}, $B^*\bar{B}^*$ with $J^{PC} = 0^{++}$, 0^{-+}, 1^{+-}, 2^{++} are about 50MeV below the corresponding $B\bar{B}^*$ and $B^*\bar{B}^*$ thresholds. No bound state, however, appears for isovectors. Isovector states receive less attraction than the states with $I = 0$, mainly because of the isospin dependence of the one-pion-exchange potential. The operator includes a factor $\tau_1.\tau_2$ whose expectation value is -3 for $I = 0$ and $+1$ for $I = 1$. Hence, if the Belle Z_b states are identified as the isovector $B\bar{B}^*$ and $B^*\bar{B}^*$ molecules, one should also have their isoscalar conterpart somewhat below (typically a few tens of MeV).

In Ref. [9], the authors also did not found any isovector $B\bar{B}^*$ bound state. The only calculation that found loosely bound states in S-wave $B\bar{B}^*$ and $B^*\bar{B}^*$ for isovectors was presented by Liu *et al.* [10], where the authors took, besides the pion, also scalar and vector mesons exchange into account, in the framework of the meson exchange model.

3. Tetraquark States

In the context of the tetraquark picture, using the chromomagnetic interaction, the authors of Ref. [14] studied the masses of the S-wave $[bq][\bar{b}\bar{q}]$ tetraquark states with $J^P = 1^+$. They found six states and two of them are consistent with the $Z_b(10610)$ and $Z_b(10650)$. However, it is important to mention that these two states are not the low-lying states in this channel. The lowest $[bq][\bar{b}\bar{q}]$ tetraquark state with $J^P = 1^+$ appears at 10167.9 MeV. This result is consistent with the findings of Ref. [21] where, using the color-magnetic interaction with the flavor symmetry breaking corrections, the $b\bar{b}q\bar{q}$ tetraquark states were predicted to be around $10.2 \sim 10.3$ GeV. These results are also consistent with the values extracted from the QCD sum rule approach [22, 23] for the $[bq][\bar{b}\bar{q}]$ tetraquark state with $J^{PC} = 1^{++}$.

The first QCD sum rule (QCDSR) calculation for the tetraquark $[bq][\bar{b}\bar{q}]$ state with $J^{PC} = 1^{++}$, which we call X_b, was performed in Ref. [22]. At the sum rule stability point and using the perturbative $\overline{MS}$-mass $m_b(m_b) = 4.24$ GeV, the authors obtained $M_{X_b} = (10250 \pm 200)$ MeV, for $\sqrt{s_0} = (10500 \pm 300)$ MeV, where s_0 is the continuum threshold. The authors of Ref. [23] have used different $J^{PC} = 1^{++}$ and $J^{PC} = 1^{+-}$ tetraquark $[bq][\bar{b}\bar{q}]$ currents. They have obtained $M_{X_b} = (10220 \pm 100)$ MeV, for $\sqrt{s_0} = (10800 \pm 100)$ MeV which is in complete agreement with the result of Ref. [22]. The authors of Ref. [24] tried to reproduce the mass of $Z_b(10610)$ using a $B\bar{B}^*$ molecular current in the QCDSR calculation. They obtained $M_{BB^*} = (10560 \pm 180)$ MeV, which they say is in agreement with the $Z_b(10610)$ mass. However, to obtain such a big mass they were forced to use a continuum threshold of $\sqrt{s_0} = (11400 \pm 200)$ MeV, which is much bigger than the values used in Refs. [22, 23]. As it was shown in Ref. [25], different currents with the same quantum numbers lead to approximately the same mass in the QCDSR approach, if the same parameters are used. Therefore, from QCDSR studies one may say that the low-lying X_b state has a mass around 10100 - 10200 MeV, which is in agreement with the results from chromomagnetic model calculations [14, 21].

It is very interesting to notice that the mass difference between the predicted X_b and $\chi_{b1}(9892)$:

$$M_{X_b} - M_{\chi_{b1}} \sim 310 \text{ MeV}, \tag{1}$$

is of the same order of magnitude of the mass difference between the $X(3872)$ and $\chi_{c1}(3510)$:

$$M_X - M_{\chi_{c1}} \sim 360 \text{ MeV}. \tag{2}$$

This kind of similarity between the c and b sector is very interesting. One can see that $M_{\Psi(2S)} - M_{\Psi(1S)} = 590$ MeV $\sim M_{\Upsilon(2S)} - M_{\Upsilon(1S)} = 560$ MeV. Therefore, the results in Eqs. (1) and (2), could be used as an evidence that a 4-quark state, similar to $X(3872)$, should exist in the b sector! This state should be searched for by the experimental groups.

We suggest that the recently observed Z_b states are not ground states of the 1^+ bottomonium 4-quark states, but excitations of a ground state with a mass around 10.2 GeV. A similar suggestion was made by Maiani, Polosa and Riquer [26], to explain the $Z^+(4430)$ as an excitation of a charged X state. Although the Babar Collaboration found no conclusive evidence of the existence of $Z^+(4430)$ [27], Belle has confirmed its observation. Using the same data sample of Ref. [6], Belle also performed a full Dalitz plot analysis [28] and has confirmed the observation of the $Z^+(4430)$ signal with a $6.4\,\sigma$ peak significance. Therefore, if one believes in the Z_b states observed by Belle, one should also believe in the existence of the $Z^+(4430)$.

4. Z_b is not X_b. Is that a question?

In [29] it was conjectured that the $X(3872)$ must have a charged partner X^+ with $J^{PC} = 1^{+-}$ with a similar mass. In [26] it was pointed out that since the mass difference

$$M_{Z^+(4430)} - M_{X^+(3872)} \sim 560 \text{ MeV} \tag{3}$$

is close to the mass difference $M_{\psi(2S)} - M_{\psi(1S)}$ given above, the $Z^+(4430)$ may well be the first radial excitation of the X^+. In a straightforward extension of this reasoning to the bottom case, we may conjecture that the $Z_b(10610)$ is also a radial excitation of an yet unmeasured X_b^+ ($J^P = 1^+$) state, such that the mass difference

$$M_{Z_b(10610)} - M_{X_b^+(10100)} \sim 510 \text{ MeV} \tag{4}$$

very close to the mass difference $M_{\Upsilon(2S)} - M_{\Upsilon(1S)} = 560$ MeV. For the sake of clarity the above mentioned numbers are displayed in Fig. 1, where we compare the charm and bottom spectra in the mass region of interest. On the left (right) we show the charm (bottom) states with their mass differences in MeV. The comparison between the two left lines with the two lines on the right emphasizes the similiraty between the conjecture made in this note, namely the existence of X_b^+ as a ground state of the measured $Z_b^+(1610)$, and the conjecture made in Ref. [26] on the existence of the charged partner X^+ of the $X(3872)$ as the ground states of the $Z^+(4430)$.

The conjecture presented above should encourage searches both in the charm and bottom sectors in the mass regions around 3870 MeV and 10200 MeV respectively. The reason for these searches would be not only to find the unobserved states shown in Fig. 1, $X^+(?)$, $X_b(?)$ and $X_b^+(?)$. In fact there are even more striking states to be detected in these mass regions. In the molecular approach, there is also attraction in the flavor exotic DD or D^*D^* and BB^* or B^*B^* channels. In the vector-vector case, the Fermi–Yang rule of G-parity [30] holds and the pion-exchange potential flips sign (for a given isospin I). Hence $B^*\bar{B}^*$ channels with repulsive

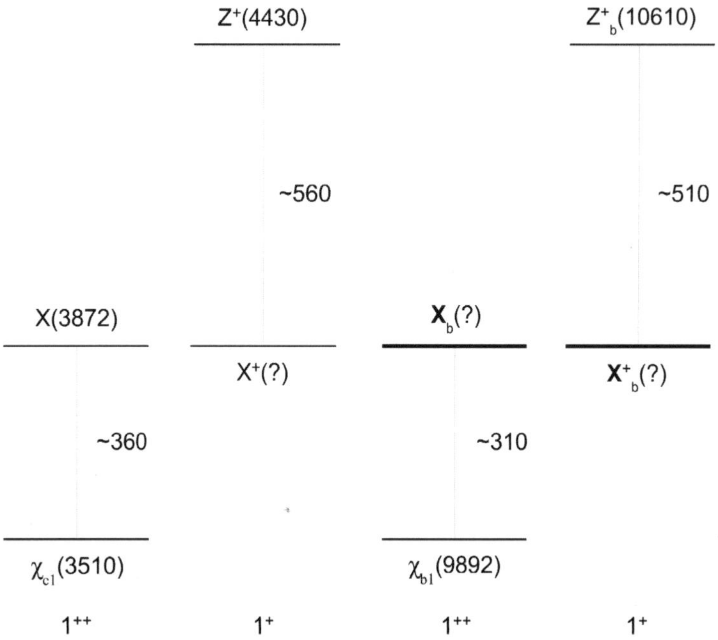

Figure 1. Charm and bottom energy levels in the mass region of interest. Masses are in MeV. On the two left columns we show the conjecture presented in [26]. The $X^+(?)$ is the proposed charged partner of the $X(3872)$. On the two right columns we show our equivalent conjecture for the bottom sector, with $X_b(?)$ and $X_b^+(?)$ being the proposed states.

interaction are transformed into B^*B^* with attraction. The pseudoscalar-vector case is more subtle. Pion-exchange induces a transition from $|1\rangle = B\bar{B}^*$ to $|2\rangle = B^*\bar{B}$, and by linear combination, the potential is attractive in one of the channels, say $|1\rangle \pm |2\rangle$. In the flavor-exotic sector, one deals with $|1'\rangle = BB^*$ and $|2'\rangle = B^*B$. The pion-exchange potential still flips signs, but this simply means that the very same attraction is observed now in the combination $|1'\rangle \mp |2'\rangle$, if it exists in the partial wave one looks at.

For tetraquark systems, constituent model calculations have always favored $(QQ\bar{q}\bar{q})$ configurations, where the pair of heavy quarks benefits from their attraction. In the threshold $(Q\bar{q}) + (Q\bar{q})$, instead, the heavy quarks interact only with light quarks, and no heavy reduced mass enters. For a review, see, e.g., [31]. This peculiarity of $(QQ\bar{q}\bar{q})$ channels is confirmed in some studies based on lattice QCD [32] and QCD sum rules [33]. Experimentally the sector with charm +2 or beauty −2 is almost virgin. Theoretically, the question is whether charm is heavy enough to make $(cc\bar{q}\bar{q})$ stable against spontaneous dissociation or b quarks are necessary. Two b or not two b, that is the question!

5. Conclusions

We have discussed the masses of some X, Y and Z states, recently observed by BaBar and Belle Collaborations. In some cases a tetraquark configuration was favored, as the $Y_b(10890)$ [34], and in some other cases a molecular configuration was favored. In the case of $Z_b^+(10610)$ we identify it as the first excitation of the tetraquark $X_b(1^{++})$, the analogue of the $X(3872)$ state in the charm sector.

Acknowledgments
This work has been partly supported by the CNRS-FAPESP program, by CNPq-Brazil and by the CNRS-IN2P3 within the project Non-perturbative QCD and Hadron Physics. We thank Frank Close, Christopher Thomas and Nils Törnqvist for a (still ongoing) lively discussion about the molecular dynamics.

References

[1] N. Brambilla, *et al.*, Eur. Phys. J. C **71**, 1534 (2011) [arXiv:1010.5827].
[2] M. Nielsen, F. S. Navarra, S. H. Lee, Phys. Rept. **497**, 41-83 (2010) [arXiv:0911.1958].
[3] S. L. Olsen, Nucl. Phys. A **827**, 53C-60C (2009) [arXiv:0901.2371].
[4] K. -F. Chen *et al.* [Belle Collaboration], Phys. Rev. D **82**, 091106 (2010) [arXiv:0810.3829].
[5] I. Adachi *et al.* [Belle Collaboration], arXiv:1105.4583.
A. Bondar *et al.* [Belle Collaboration] arXiv:1110.2251.
[6] S. K. Choi *et al.* [Belle Collaboration], Phys. Rev. Lett. **100**, 142001 (2008).
[7] A. E. Bondar, A. Garmash, A. I. Milstein, R. Mizuk and M. B. Voloshin, arXiv:1105.4473.
[8] Y. Yang, J. Ping, C. Deng and H. S. Zong, arXiv:1105.5935.
[9] J. Nieves and M. P. Valderrama, arXiv:1106.0600.
[10] Z.-F. Sun, J. He, X. Liu, Z.-G. Luo and S.-L. Zhu, arXiv:1106.2968.
[11] M. Cleven, F.-K. Guo, C. Hanhar and U.-G. Meissner, arXiv:1107.0254.
[12] D. V. Bugg, Europhys. Lett. **96**, 11002 (2011). [arXiv:1105.5492].
[13] I. V. Danilkin, V. D. Orlovsky and Yu. A. Simonov, arXiv:1106.1552.
[14] T. Guo, L. Cao, M. Z. Zhou and H. Chen, arXiv:1106.2284.
[15] M. B. Voloshin and L. B. Okun, JETP Lett. **23**, 333 (1976) [Pisma Zh. Eksp. Teor. Fiz. **23**, 369 (1976)].
[16] A. De Rujula, H. Georgi and S. L. Glashow, Phys. Rev. Lett. **38**, 317 (1977).
[17] A. Le Yaouanc, L. Oliver, O. Pene and J. C. Raynal, Phys. Lett. B **71**, 397 (1977); **72**, 57 (1977);
E. Eichten, K. Gottfried, T. Kinoshita, K. D. Lane and T. M. Yan, Phys. Rev. D **21**, 203 (1980).
[18] N. A. Tornqvist, Z. Phys. C **61**, 525 (1994).
[19] T. E. O. Ericson and G. Karl, Phys. Lett. B **309**, 426 (1993).
[20] A. V. Manohar and M. B. Wise, Nucl. Phys. B **399**, 17 (1993).
[21] Y. Cui, X. L. Chen, W. Z. Deng and S. L. Zhu, High Energy Phys. Nucl. Phys. **31**, 7 (2007).
[22] R.D. Matheus, S. Narison, M. Nielsen, J.M. Richard, Phys. Rev. **D75**, 014005 (2007).
[23] W. Chen and S. L. Zhu, Phys. Rev. D **83**, 034010 (2011).
[24] J. R. Zhang, M. Zhong and M. Q. Huang, Phys. Lett. **B704**, 312-315 (2011). [arXiv:1105.5472].
[25] S. Narison, F.S. Navarra, M. Nielsen, Phys. Rev. **D83**, 016004 (2011).
[26] L. Maiani, A. D. Polosa and V. Riquer, arXiv:0708.3997, and New Journal of Physics, **10** 073004 (2008).
[27] B. Aubert *et al.* [BaBar Collaboration], Phys. Rev. D **79**, 112001 (2009).
[28] R. Mizuk *et al.* [Belle Collaboration], Phys. Rev. D **80**, 031104(R) (2009).
[29] L. Maiani, F. Piccinini, A. D. Polosa and V. Riquer, Phys. Rev. D **71**, 014028 (2005).
[30] It was much used in the time of baryonium to relate the nucleon–antinucleon interaction to the nucleon-nucleon one.
[31] J. Vijande, A. Valcarce, J. M. Richard and N. Barnea, Few Body Syst. **45**, 99 (2009).
[32] C. Michael and P. Pennanen [UKQCD Collaboration], Phys. Rev. D **60**, 054012 (1999). M. Wagner [ETM Collaboration], arXiv:1103.5147.
[33] F. S. Navarra, M. Nielsen and S. H. Lee, Phys. Lett. B **649**, 166 (2007).
[34] R. M. Albuquerque, M. Nielsen, R. R. da Silva, arXiv:1110.2113.

Journal of Physics: Conference Series **348** (2012) 012008 doi:10.1088/1742-6596/348/1/012008

Double charm states in QCD sum rules

Fernando S. Navarra[1] **and Marina Nielsen**[1]

[1]Instituto de Física, Universidade de São Paulo,
Caixa Postal 66318, 05389-970 São Paulo, SP, Brazil

E-mail: navarra@if.usp.br, mnielsen@if.usp.br

Abstract.
In this note we revisit and improve the calculation of the T_{cc} mass using the double ratio of QCD sum rules (QCDSR), assuming that this state is described by a molecular current. In a previous work we used a tetraquark current. We conclude that with both currents we arrive at nearly the same results for the T_{cc} mass, which turns out to be close to the mass of the $X(3872)$.

1. Introduction

Recently we have learned that there are hadronic states whose quantum numbers and main properties cannot be explained by a simple quark-antiquark or three-quark configuration. These states are called exotic states. The most well studied state is the $X(3872)$ resonance (assumed to be an 1^{++} axial vector meson), which was discovered by BELLE in B-decays and confirmed by BABAR, CDF and D0 (for a recent review see [1]). Its most popular picture consists of a molecular configuration, $D\bar{D}^* + \bar{D}D^*$. Here we focus on another exotic meson, which has similar structure, simliar quark content and similar mass: the four-quark state $(QQ\bar{u}\bar{d})$ with quantum numbers $I = 0,\ J = 1$ and $P = +1$ which, following Ref.[2], we call T_{QQ}, where Q stands for c or b. As already noted previously [2, 3], the T_{bb} and T_{cc} states with $J^P = 1^+$ cannot split into a pair of two $\bar{B}$ or two D mesons. If their masses are below the $\bar{B}\bar{B}^*$ or $DD\pi$ thresholds, these decays are also forbidden. As a result, T_{QQ} becomes stable with respect to the strong interaction, and must decay radiatively, or even weakly if the mass becomes lower than the threshold made of two pseudoscalar mesons.

In constituent models with a flavor-independent central potential, the stability of tetraquark $(QQ\bar{q}\bar{q})$ configurations comes from a favorable effect when the charge-conjugation symmetry is broken. This is the same mechanism by which, in QED, the loosely bound positronium molecule evolves into the very stable hydrogen molecule. It has been observed that in the large m_Q limit, the light degrees of freedom cannot resolve the closely bound QQ system. This results in bound states similar to the $\bar{\Lambda}_Q$ states, with QQ playing the role of the heavy antiquark [4]. The $(QQ\bar{q}\bar{q})$ states have been studied using a variety of potential models [3, 5, 6]. The corresponding four-body problem is very delicate, as discussed in [7]. All authors agree that such states become bound when the quark over the antiquark mass ratio becomes sufficiently large. Detailed four-body calculations, using a pairwise central potential supplemented by a chromomagnetic interaction, indicate that T_{bb} is rather well bound, and T_{cc} possibly bound by a few MeV below DD^*. For instance, the prediction of Ref. [2] is, in units of MeV:

$$M_{T_{cc}} = 3876 \sim 3905\ , \quad M_{T_{bb}} = 10519 \sim 10651\ . \tag{1}$$

Journal of Physics: Conference Series **348** (2012) 012008 doi:10.1088/1742-6596/348/1/012008

A non-pairwise confinement has also been considered [8], inspired by the large coupling regime of QCD, where it is shown [9] that it is more favorable to build stable tetraquarks. In this improved quark model, as well as in conventional quark models, it is found that $(QQ\bar{q}\bar{q})$ has an energy lower than $(Q\bar{Q}q\bar{q})$. Another variant was considered in [10], with a chiral potential model, which includes meson-exchange forces between quarks, instead of the chromomagnetic interaction.

The existence of a $D\bar{D}^* + \bar{D}D^*$ molecule was predicted in Ref. [11] on the basis of the pion-exchange dynamics. A comprehensive discussion on meson molecules can be found in [12]. Here, the pion is exchanged between the hadrons, as in the Yukawa theory of nuclear forces. The $DD^*\pi$ and $\bar{D}\bar{D}^*\pi$ vertices are identical, as well as the $D^*D^*\pi$ and $\bar{D}^*\bar{D}^*\pi$ ones. There is only an overall change of sign, due to the G-parity of the pion. Therefore, if the pion-exchange dynamics is able to bind the $D\bar{D}^* + \bar{D}D^*$ molecule, the same is true for the DD^* molecule. The difference between these two states can only come from the short-range part of the interaction, which is accurately described by QCDSR.

In what follows we discuss the mass determination of the T_{QQ} and the ways to improve it in the framework of QCDSR. We explore the connection between the $X(3872)$ and the T_{QQ} and argue that the existence of the first implies the existence of the second.

2. T_{QQ} in QCD sum rules

The first study of tetraquarks with two heavy quarks within QCD sum rules was done in [13], where these states were represented by diquark-antidiquark currents. This study was revisited and improved in [14]. Here we compare in detail the $(QQ\bar{q}\bar{q})$ and $(Q\bar{Q}q\bar{q})$ configurations. Such a comparison was also performed in Ref. [15], where the authors have studied heavy tetraquarks using a crude color-magnetic interaction, with flavor symmetry breaking corrections. In [13] it was found that (in units of GeV):

$$M_{T_{cc}} = 4.0 \pm 0.2 \ , \qquad M_{T_{bb}} = 10.2 \pm 0.3 \ . \tag{2}$$

The short-range part of the interaction can be tested by the QCD sum rules approach. Therefore, we can study the ratio of the masses of the T_{cc} and $X(3872)$ states, by using the double ratios of sum rules (DRSR), which is widely applied for accurate determinations of the ratios of couplings and masses and form factors. This accuracy is reached due to partial cancellations of the systematics of the method and of the QCD corrections in the DRSR. More recently, the DRSR was used to study different possible currents for the $X(3872)$ [16]. It was found that (within the accuracy of the method) the different structures ($\bar{3}-3$ and $\bar{6}-6$ tetraquarks and $D\bar{D}^* + \bar{D}D^*$ molecule) lead to the same prediction for the mass. The two-point functions of the $X(3872)$ (assumed to be an 1^{++} axial vector meson) and the T_{cc} (assumed to be a $J^P = 1^+$ state) is defined as:

$$\begin{aligned} \Pi_i^{\mu\nu}(q) &\equiv i\int d^4x \ e^{iq.x}\langle 0|T[j_i^\mu(x)j_i^{\nu\dagger}(0)]|0\rangle \\ &= -\Pi_{1i}(q^2)(g^{\mu\nu} - \frac{q^\mu q^\nu}{q^2}) + \Pi_{0i}(q^2)\frac{q^\mu q^\nu}{q^2}, \end{aligned} \tag{3}$$

where $i = X,\ T_{cc}$. The two invariants, Π_1 and Π_0, appearing in Eq. (3) are independent and have respectively the quantum numbers of the spin 1 and 0 mesons.

We assume that the $X(3872)$ and T_{cc} states are described by the molecular currents:

$$j_X^\mu(x) = \bar{q}_a(x)\gamma_5 c_a(x)\bar{c}_b(x)\gamma^\mu q_b(x) - \bar{q}_a(x)\gamma^\mu c_a(x)\bar{c}_b(x)\gamma_5 q_b(x) \ , \tag{4}$$

and

$$j_{T_{cc}}^\mu(x) = \bar{q}_a(x)\gamma_5 c_a(x)\bar{q}_b(x)\gamma^\mu c_b(x) \tag{5}$$

where a and b are color indices. The correlation function, Π_{1i} in Eq. (3), can be written in terms of a dispersion relation:

$$\Pi_{1i}(q^2) = \int_{4m_c^2}^{\infty} ds \frac{\rho_i(s)}{s - q^2} + \cdots , \tag{6}$$

where $\pi\rho_i(s) \equiv \text{Im}[\Pi_{1i}(s)]$ is the spectral function.

The correlation function in Eq. (3) is evaluated in two ways: using the operator product expansion (OPE) and using the information from hadronic phenomenology. In the OPE side we work at leading order of perturbation theory in α_s, and we consider the contributions from condensates up to dimension six. In the phenomenological side, the correlation function is estimated by inserting intermediate states for the X and T_{cc} states via their couplings λ_i to the molecular currents, $\langle 0|j_i^\mu|M_i\rangle = \lambda_i \epsilon^\mu$, where $M_i \equiv X,\ T_{cc},\ j_i^\mu$ are the currents in Eqs. (4) and (5).

Using the Ansatz: "one resonance plus QCD continuum", where the QCD continuum comes from the discontinuity of the QCD diagrams from a continuum threshold t_c, the phenomenological side of Eq. (3) can be written as:

$$\Pi_{\mu\nu}^{phen}(q^2) = \frac{\lambda_i^2}{M_i^2 - q^2}\left(-g_{\mu\nu} + \frac{q_\mu q_\nu}{M_i^2}\right) + \cdots , \tag{7}$$

where the Lorentz structure $g_{\mu\nu}$ projects out the 1^+ state. The dots denote higher axial-vector resonance contributions that will be parametrized, as usual, by the QCD continuum. After making a Borel transform on both sides, and transferring the continuum contribution to the QCD side, the moment sum rule and its ratio read:

$$\begin{aligned} \mathcal{F}_i(\tau) &\equiv \lambda_i^2 e^{-M_i^2\tau} = \int_{4m_c^2}^{t_c} ds\ e^{-s\tau}\ \rho_i(s) \\ \mathcal{R}_i(\tau) &\equiv -\frac{d}{d\tau}\log \mathcal{F}_i(\tau) \simeq M_i^2 , \end{aligned} \tag{8}$$

where $\tau \equiv 1/M^2$ is the sum rule variable with M being the Borel mass. In the following, we shall work with the DRSR:

$$r_{T_{cc}/X} \equiv \sqrt{\frac{\mathcal{R}_{T_{cc}}}{\mathcal{R}_X}} \simeq \frac{M_{T_{cc}}}{M_X}. \tag{9}$$

Using QCD sum rules, one can usually estimate the mass of the X-meson, from the ratio $\mathcal{R}_X$ in Eq. (8), which is related to the spectral densities obtained from the current (4). A tetraquark current for the $X(3872)$ was used in Ref. [17] and its mass was estimated to be:

$$M_X \simeq \sqrt{\mathcal{R}_X} = (3925 \pm 127)\ \text{MeV} . \tag{10}$$

Although the uncertainty in Eq. (10) is still large, considering the fact that this result was obtained in a Borel region where there is pole dominance and OPE convergence, one can say that the QCD sum rules support the existence of such a state and that the value obtained for M_X is in reasonable agreement with the experimental value $M_X = (3872.2 \pm 0.8)$ MeV .

We now study the DRSR of the T_{cc}/X defined in Eq. (9). In Fig. 1, we show the τ-dependence of the ratio for $\sqrt{t_c} = 4.15$ GeV and for two values of $m_c = 1.23$ GeV and 1.47 GeV. From Fig. 1 one can see that there is a τ-stability around $\tau \simeq 0.4$ GeV^{-2} and for this value of τ, we get:

$$r_{T_{cc}/X} = 1.00 \pm 0.01 . \tag{11}$$

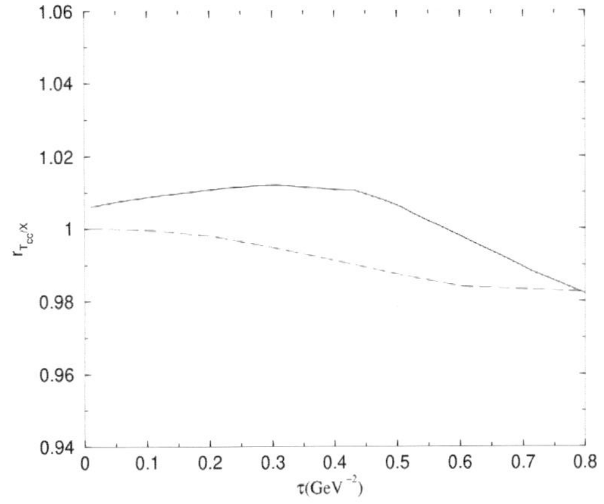

Figure 1. The double ratio $r_{T_{cc}/X}$ defined in Eq. (9) as a function of τ for $\sqrt{t_c} = 4.15$ GeV and for two values of $m_c = 1.23$ (solid line) and 1.47 GeV (dashed line).

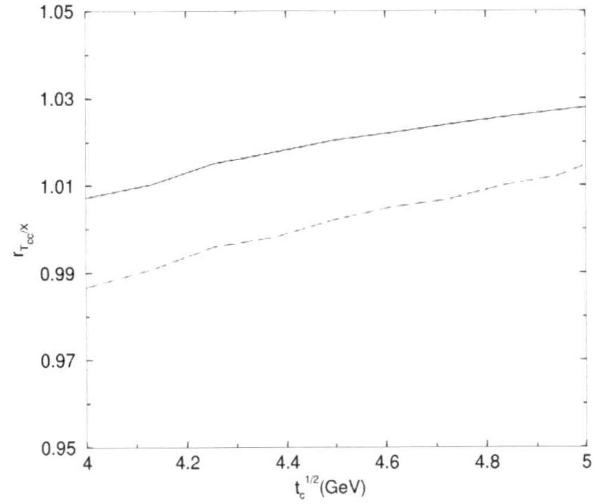

Figure 2. The double ratio $r_{T_{cc}/X}$ as a function of t_c for $\tau = 0.4$ GeV^{-2} and for two values of $m_c = 1.23$ (solid line) and 1.47 GeV (dashed line).

In Fig. 2, we show the t_c-dependence of the ratio for $\tau = 0.4$ GeV^{-2} and for two values of $m_c = 1.23$ GeV and 1.47 GeV. From this figure one can see that the ratio increases with t_c. However, considering the large range of t_c presented in the figure, the ratio does not differ more than 3% from 1.

Our analysis has shown that the $D\bar{D}^* + c.c.$ and DD^* currents lead to the same mass predictions within the accuracy of the approach. The accuracy of the DRSR is bigger than the normal QCDSR because the DRSR are less sensitive to the exact value and definition of the heavy quark mass and to the QCD continuum contributions . As mentioned before, this accuracy is reached due to partial cancellations of the systematics of the method and of the QCD corrections in the DRSR. Therefore, if the observed $X(3872)$ is a molecular $D\bar{D}^* + c.c.$ state its molecular cousin DD^* should also be a bound state. Its mass can be obtained by using the experimental mass for the $X(3872)$ in Eq. (11):

$$M_{T_{cc}} = (3872.2 \pm 39.5) \text{ MeV} . \tag{12}$$

Using the same interpolating field in Eqs. (4) and (5) with the charm quark replaced by the bottom one, we can analyse the DRSR:

$$r_{T_{bb}/X_b} \equiv \sqrt{\frac{\mathcal{R}_{T_{bb}}}{\mathcal{R}_{X_b}}} \simeq \frac{M_{T_{bb}}}{M_{X_b}} \ . \tag{13}$$

We obtain:

$$r_{T_{bb}/X_b} = 1.00 \ . \tag{14}$$

Therefore, we can predict the degeneracy between the masses of the T_{bb} and of the X_b.

3. Conclusions

In this note we have presented a refinement of the determination of the T_{cc} mass. In [13], using a tetraquartk current we arrived at $M_{T_{cc}} = 4.0 \pm 0.2$ GeV. Here, using a molecular current and the DRSR method we arrive at $M_{T_{cc}} = 3.87 \pm 0.04$ GeV. The DRSR method relates two particles with the same quantum numbers and, due to the cancellations mentioned above, allows for a more precise determination of the mass. In the present case the associated particle is the well established $X(3872)$. Within the uncertainties the new determination of the T_{cc} mass agrees with the previous one. The sum rules for the $X(3872)$ and for the T_{cc} have the same (good) quality. This suggests that the existence of the former implies the existence of the latter.

From the microscopic point of view, QCDSR describes accurately short range interactions. If both the T_{cc} and the $X(3872)$ are meson molecules we should also include long range interactions, i.e., meson exchanges. If the dominant long range process is the pion exchange between the corresponding mesons, in the present case it would be exactly the same for the two molecules in question. We can then conclude that if the observed $X(3872)$ meson is a $D\bar{D}^*+$ molecule, then the DD^* molecule, associated to the T_{cc}, should also exist with approximately the same mass. An extension of the analysis to the b-quark case leads to the same conclusion. This question will certainly be experimentally studied at the LHC in the near future.

Acknowledgments

This work has been partly supported by FAPESP and by CNPq. The authors are deeply grateful to the organizaers of the TROIA 11 Conference for their warm hospitality and for many stimulating discussions: G. Erkol, A. Kucukarslan, A. Ozpineci, K. Utku Can.

References

[1] M. Nielsen, F. S. Navarra, S. H. Lee, Phys. Rep. **497**, 41 (2010).
[2] D. Janc and M. Rosina, Few Body Sust. **35**, 175 (2004).
[3] S. Zouzou, B. Silvestre-Brac, C. Gignoux and J.M. Richard, Z. Phys. **C30**, 457 (1986).
[4] H. J. Lipkin, Phys. Lett. **B172**, 242 (1986); C.-K. Chow, Phys. Rev. **D51**, 3541 (1995).
[5] J. Vijande, A. Valcarce and K. Tsushima, Phys. Rev. **D74**, 054018 (2006).
[6] D.M. Brink and Fl. Stancu, Phys. Rev. **D57**, 6778 (1998).
[7] J. Vijande, A. Valcarce and N. Barnea, Phys. Rev. D **79**, 074010 (2009).
[8] J. Carlson and V. R. Pandharipande, Phys. Rev. D **43**, 1652 (1991).
[9] J. Vijande, A. Valcarce and J. M. Richard, Phys. Rev. D **76**, 114013 (2007).
[10] S. Pepin, Fl. Stancu, M. Genovese and J.M. Richard, Phys. Lett. **B393**, 119 (1997).
[11] N.A. Törnqvist, Z. Phys. **C61**, 525 (1994).
[12] E. S. Swanson, Phys. Rept. **429**, 243 (2006).
[13] F.S. Navarra, M. Nielsen and S.H. Lee, Phys. Lett. **B649**, 166 (2007).
[14] J. M. Dias, S. Narison, F. S. Navarra, M. Nielsen, J. -M. Richard, Phys. Lett. B **703**, 274 (2011).
[15] Y. Cui, X.-L. Chen, W.-Z. Deng and S.-L. Zhu, High Energy Phys. Nucl. Phys. **31**, 7 (2007).
[16] S. Narison, F.S. Navarra, M. Nielsen, Phys. Rev. **D83**, 016004 (2011).
[17] R.D. Matheus, S. Narison, M. Nielsen, J.M. Richard, Phys. Rev. **D75**, 014005 (2007).

Unitary symmetry pattern of the QCD sum rules

T.Aliev[1], A.Ozpineci[1] and V.Zamiralov[2]

[1] Physical Department, Middle East Technical University, Ankara, Turkey

[2] D.V.Skobeltsyn Institute of Nuclear Physics, Moscow State University, Moscow, Russia

E-mail: taliev@metu.edu.tr,ozpineci@metu.edu.tr,zamir@depni.sinp.msu.ru

Abstract. Relations between magnetic moments of Σ^0 and Λ baryons are discussed. Physical meaning of the F- and D- couplings in $SU(3)$ is established. The obtained results are generalized to the QCD sum rules which yield unitary pattern with the characteristic F and D structures.

QCD sum rules introduced in [1] are now one of the most reliable methods of analyzing strong interaction. Dozens of works have described couplings of baryons to γ, pseudoscalar and vector mesons, their weak decay constants etc. We have shown that unitary symmetry plays essential role in relating between them various couplings through well-known F- and D-type structures reducing number of independent correlation functions to minimum. We would show here in what way unitary pattern of the QCD sum rules arrives.

Let us begin with the discussion of the baryon octet $1/2^+$ in $SU(3)$.

There are 3 different reductions of the octet along the $SU(2)$.

The 1st one is with Λ being the isosinglet and Σ the isotriplet, isomultiplets formed along the value of Y. The 2nd one is with 11 component $\Sigma^0/\sqrt{2}+\Lambda/\sqrt{6} \equiv -2\Lambda_{us}/\sqrt{6}$ being the U-singlet and n, Ξ^0, $\Sigma^0/2+\sqrt{3}\Lambda/2 \equiv \Sigma_{us}$ form U-triplet, $SU(2)_U$-multiplets having the same charge Q. The 3rd possibility is given by 22-component $-\Sigma^0/\sqrt{2}+\Lambda/\sqrt{6} \equiv -2\Lambda_{ds}/\sqrt{6}$ being the V-singlet and p, Ξ^-, $-\Sigma^0/2+\sqrt{3}\Lambda/2 \equiv \Sigma_{ds}$ form V-triplet, $SU(2)_V$-multiplets having the same $Q-Y$.

For any operator $\mathcal{O}$ the following relations are valid between the matrix elements:

$$2\langle\Sigma_{us}|\mathcal{O}|\Sigma_{us}\rangle + 2\langle\Sigma_{us}|\mathcal{O}|\Sigma_{us}\rangle - \langle\Sigma^0|\mathcal{O}|\Sigma^0\rangle = 3\langle\Lambda|\mathcal{O}|\Lambda\rangle, \quad (1)$$

$$2\langle\Lambda_{us}|\mathcal{O}|\Lambda_{us}\rangle + 2\langle\Lambda_{ds}|\mathcal{O}|\Lambda_{ds}\rangle - \langle\Lambda|\mathcal{O}|\Lambda\rangle = 3\langle\Sigma^0|\mathcal{O}|\Sigma^0\rangle.$$

To show its applicability we consider an example of magnetic moments of octet baryons. In NRQM one has $\mu_\Lambda = \mu_s$ [2]. Now instead of proceeding quark framework calculations we just use relations (1) to obtain μ_{Σ^0} from μ_Λ : $2\mu_{\Lambda_{ds}} + 2\mu_{\Lambda_{us}} - \mu_\Lambda = 2\mu_d + 2\mu_u - \mu_s = 3\mu_{\Sigma^0}$.
Let us now assume that photon (or ρ^0, ω, ϕ) interacts in a different way with two quarks of similar flavour of the Σ-like baryon $B(qq, q')$ and with a single quark q'. As an example let the magnetic moment operator has the form $\hat{e}_q \hat{\omega}_q \hat{\sigma}_z^q$, where new operator $\hat{\omega}_q$ just differs between a single q' quark and a diquark $(q_\uparrow q_\uparrow)$ or $(q_\uparrow q_\downarrow)$ through the matrix elements

$$< q_\uparrow q_\uparrow, q'_\downarrow|\hat{\omega}_q|q_\uparrow q_\uparrow, q'_\downarrow >= w_{\uparrow\uparrow}, \quad < q_\uparrow q_\downarrow, q'_\uparrow|\hat{\omega}_q|q_\uparrow q_\downarrow, q'_\uparrow >= w_{\uparrow\downarrow},$$
$$< q_\uparrow q_\uparrow, q'_\downarrow|\hat{\omega}_{q'}|q_\uparrow q_\uparrow, q'_\downarrow >= v_{\uparrow\uparrow}, \quad < q_\uparrow q_\downarrow, q'_\uparrow|\hat{\omega}_{q'}|q_\uparrow q_\downarrow, q'_\uparrow >= v_{\uparrow\downarrow}. \tag{2}$$

$$\mu_p = \sum\nolimits_{q=u,d} < p_\uparrow| \quad \hat{e}_q \hat{\omega}_q \hat{\sigma}_z^q \quad |p_\uparrow >=$$
$$= \frac{1}{6} < 2u_1u_1d_2 - u_1d_1u_2 - d_1u_1u_2| \quad \hat{e}_q \hat{\omega}_q \sigma_z^q \quad |2u_1u_1d_2 - u_1d_1u_2 - d_1u_1u_2 >=$$
$$= \frac{4}{3} e_u w_{\uparrow\uparrow} - \frac{1}{3} e_d (2v_{\uparrow\uparrow} - v_{\uparrow\downarrow}) \equiv e_u \cdot 2\mu_F + e_d \cdot (\mu_F - \mu_D) = \mu_F + \frac{1}{3}\mu_D$$

with $w_{\uparrow\uparrow} = 3F/2, (2v_{\uparrow\uparrow} - v_{\uparrow\downarrow})/3 = (\mu_D - \mu_F)$. Note that m.e. $w_{\uparrow\downarrow}$ drops out and does not contribute to magnetic moments of Σ-like baryons. But it contributes in axial-vector currents and we should put $w_{\uparrow\downarrow} = D$ to come to SU(3) result. It is worth noting that the assumption $w_{\uparrow\uparrow} = w_{\uparrow\downarrow}$ yields $F/D = 2/3$! The main results are:

- The F coupling is related to the interaction of the γ (or π^0, η, ρ^0, ω, ϕ) with 'diquark' composed of two quarks of (almost) equal flavour and equal spin projections
- The $(F - D)$ is related to the interaction of the γ (or π^0, η, ρ^0, ω, ϕ) with the single quark
- Λ quantities can be obtained through Eq.(1).

Coupling constant of $M = \pi^0$, η, ρ^0, ω, ϕ or photon to Σ^0 containing two quarks u, d in a diquark state and a single quark s would have the form

$$g_{M\Sigma\Sigma} = (g_{Muu} + g_{Mdd})F + g_{Mss}(F - D). \tag{3}$$

In order to obtain coupling to Λ hyperon we use the Eq.(1). Results are in accord with the SU(3) ones. This reasoning can be transferred to the QCD sum rules.
The starting point would be polarization operator (correlator)[1],[3] for hyperons Σ^0 and Λ

$$\Pi^{\Sigma^0,\Lambda} = i \int d^4x e^{ipx} \langle 0|T\{\eta^{\Sigma^0,\Lambda}(x), \eta^{\Sigma^0,\Lambda}(0)\}|V\rangle, \tag{4}$$

where isovector and isocalar interpolating currents [3] could be chosen as
$\eta^{\Sigma^0} = \frac{\epsilon_{abc}}{2}[u^{aT}Cs^b\gamma_5 d^c - d^{aT}Cs^b\gamma_5 u^c - (Cs^b\gamma_5 \to C\gamma_5 s^b)]$, $\sqrt{3}\eta^\Lambda = \tilde{\eta}^{\Sigma^0(u\leftrightarrow s)} - \tilde{\eta}^{\Sigma^0(d\leftrightarrow s)}$

where a, b, c are the color indices, C is charge conjugation matrix, while $V = \rho^0, \omega, \phi, \gamma$.
The idea of the **QCD sum rules**[1],[3] is as follows. Polarization operator is calculated in two different ways: (1) Upon using some phenomenological pole model saturated by baryon poles and resonances plus high energy contributions; (2) Upon performing Wilson operator product expansion (OPE) and calculating quark diagrams with insertions of non-zero vacuum expectation values. Equating (1)=(2) and performing Borel transformation one arrives at desired sum rule. Following the reasoning of the previous examples we construct relations which connect QCD sum rules for Σ^0 hyperon with those for Λ making explicit F,D pattern of them. For this purpose we write U-spin and V-spin quantities for the interpolating currents:

$$-2\tilde{\eta}^{\Lambda(d\leftrightarrow s)} = \sqrt{3}\eta^{\Sigma^0} + \eta^{\Lambda}, \quad 2\tilde{\eta}^{\Sigma^0(d\leftrightarrow s)} = \eta^{\Sigma^0} - \sqrt{3}\eta^{\Lambda},$$
$$2\tilde{\eta}^{\Lambda(u\leftrightarrow s)} = \sqrt{3}\eta^{\Sigma^0} - \eta^{\Lambda}, \quad 2\tilde{\eta}^{\Sigma^0(u\leftrightarrow s)} = \eta^{\Sigma^0} + \sqrt{3}\eta^{\Lambda}.$$

With these equations two-point functions of the Eq.(4) for Σ^0 and Λ are related as

$$2\cdot[\tilde{\Pi}^{\Sigma^0(d\leftrightarrow s)} + \tilde{\Pi}^{\Sigma^0(u\leftrightarrow s)}] - \Pi^{\Sigma^0} = 3\Pi^{\Lambda}, \quad 2\cdot[\tilde{\Pi}^{\Lambda(d\leftrightarrow s)} + \Pi^{\tilde{\Lambda}(u\leftrightarrow s)}] - \Pi^{\Lambda} = 3\Pi^{\Sigma^0}.$$

We can write now in full analogy to the Eq.(3)

$$\Pi^{\Sigma^0}(u,d,s) = (g_{uuV} + g_{ddV})\Pi_F(u,d,s) + g_{ssV}\Pi_{(F-D)}(u,d,s).$$

Couplings with strange mesons are obtained through auxiliary relations (below V is vector nonet)

$$\Pi^{\Sigma^0_{ds}\Sigma^-\rho^+}(u,d,s) = -\Pi^{\Sigma^0\Sigma^-\rho^+}(u,d,s) + \sqrt{3}\Pi^{\Lambda\Sigma^-\rho^+}(u,d,s).$$

By new $d \leftrightarrow s$ on the left we obtain $\Pi^{\Sigma^0\Xi^-K^{*+}}(u,d,s)$ in terms of Π_F and $\Pi_{(F-D)}$. In this way we arrive at all the correlators $\Pi^{BB'V}$ [4]. We also proved our relations for [5] and [6]. Thus,

- Relations between the magnetic moments of Σ^0 and Λ baryons are written in unitary symmetry and quark model
- Physical meaning of the F- and D- couplings of $SU(3)$ baryon currents is established
- The obtained results are generalized to the QCD sum rules
- QCD sum rules for the strong meson-baryon coupling constants yield unitary pattern with the characteristic F and D structures.

References

[1] Shifman M A, Vainshtein V I and Zakharov V I 1979 *Nucl.Phys.* B **147** 385
[2] Morpurgo G 1965 *Physics* **2** 95
[3] Ioffe B L and Smilga A 1984 *Nucl.Phys.* B **232** 109; Balitsky I I and Yung A V 1983 *Phys. Lett.* B **129** 328
[4] Aliev T M, Özpineci A, Yakovlev S and Zamiralov V 2006 *Phys. Rev.* D **D74** 116001;
Aliev T M, Özpineci A, Savci M and Zamiralov V 2009 *Phys. Rev.* D **D80** 016010
[5] Aliev T M, Özpineci A and Savci M 2000 *Phys. Rev.* D **62** 053012; 2000 *Nucl. Phys.* A **A678** 443
[6] Jin Y and Furnstahl R J 1994 *Phys. Rev.* C **49** 1190; Jin Y and Nielsen M 1995 *Phys. Rev.* C **51** 347

Analysis of $D_s^*DK_1$ ($B_s^*BK_1$) coupling constants within QCD sum rules

N Yinelek*[1], J Y Süngü[1], H Sundu[1] and K Azizi[2]

[1] Department of Physics, Kocaeli University, 41380 Izmit, Turkey

[2] Department of Physics, Doğuş University, Acıbadem-Kadıköy, 34722 Istanbul, Turkey

E-mail: neseyinelek@gmail.com

Abstract. We compute the strong coupling constants for $D_s^*DK_1$ and $B_s^*BK_1$ vertices within the QCD sum rules. The coupling constants are calculated for both $D(B)$ and K_1 off-shell mesons.

1. Introduction

The strong coupling constants among particles play an important role in providing knowledge of their strong interactions. These couplings take place in low energies, so to calculate them, we cannot apply the perturbation theory. In this study, we use the QCD sum rules approach [1] as one of the most powerful and applicable tools to calculate these coupling constants.

2. QCD Sum Rules for $D_s^*DK_1$ ($B_s^*BK_1$) Coupling Constants

To compute the $g_{D_s^*DK_1}$ and $g_{B_s^*BK_1}$ coupling form factors for the $D_s^*DK_1$ ($B_s^*BK_1$) vertices, we start with the following three-point correlation function for the $D(B)$ off-shell state:

$$\Pi_{\mu\nu(\nu\nu')}^{D(B)} = i^2 \int d^4x\, d^4y\, e^{ip'.x} e^{iq.y} \langle 0 | T\left(j_{\nu(\nu\nu')}^{K_1}(x)\, j^{D(B)}(y)\, j_\mu^{D_s^*(B_s^*)\dagger}(0) \right) | 0 \rangle \qquad (2.1)$$

Here T is the time ordering product, p' and q are the momenta of the final on-shell state and the off-shell state, respectively. The momentum of the initial state is defined as $p = p' + q$. Here, we should mention that the axial K_1 meson couples into two interpolating currents $j_\nu^{K_1}(x) = \bar{s}(x)\gamma_\nu\gamma_5 u(x)$ and $j_{\nu\nu'}^{K_1}(x) = \bar{s}(x)\sigma_{\nu\nu'}\gamma_5 u(x)$. The two remaining interpolating currents are given as $j^{D(B)}(x) = \bar{u}(x)\gamma_5 c(b)(x)$ and $j_\mu^{D_s^*(B_s^*)} = \bar{s}(x)\gamma_\mu c(b)(x)$. The three-point correlation function can be calculated in two different ways, namely QCD and physical sides [1]. The correlation function in QCD side is written as the sum of the perturbative and non-perturbative parts. The perturbative part is written in terms of double dispersion integral as:

$$\Pi_{\mu\nu(\nu\nu')}^{pert.} = -\frac{1}{4\pi^2} \int ds' \int ds \frac{\rho_{\mu\nu(\nu\nu')}(s, s', q^2)}{(s - p^2)(s' - p'^2)} + \text{subtraction terms} \qquad (2.2)$$

here $\rho_{\mu\nu(\nu\nu')}$ is spectral density. We calculate the bare loop diagram to get the spectral density and quark-quark and quark-gluon condensate diagrams to obtain the non-perturbative contributions [2]. As a result, we obtain:

$$\begin{aligned}
\rho_1^{D(B)}(s,s',q^2) &= -2N_c I_0(s,s',q^2)\big[2m_s^3+2m_s^2m_u-2m_{c(b)}m_s(m_s+m_u)+m_s(s+s'-q^2)+4A(m_{c(b)}-m_u)\\
&\quad +B\big(2m_u s+m_{c(b)}(q^2-s-s')+m_s(-q^2+3s+s')\big)\\
&\quad +C\big(-2m_{c(b)}s'+m_u(-q^2+s+s')+m_s(-q^2+s+3s')\big)\big],\\
\rho_2^{D(B)}(s,s',q^2) &= 4N_c I_0(s,s',q^2)\big[2A+(m_{c(b)}m_s-m_s^2)-(B+2G)s+H\,(q^2-s-s')\\
&\quad +C\big(m_{c(b)}(m_s+m_u)-m_s(m_s+m_u)+q^2-s-s'\big)\big],
\end{aligned} \tag{2.3}$$

where $N_c=3$ is the colour number and

$$\begin{aligned}
\Pi^{D(B)}_{nonpert1} &= -\langle\bar{s}s\rangle\left[\frac{m_0^2}{24}\left(\frac{m_{c(b)}^2-q^2}{r^2r'}+\frac{2}{rr'}+\frac{1}{r^2}+\frac{1}{r'^2}+\frac{m_{c(b)}^2-q^2}{rr'^2}\right)+\frac{q^2+2m_um_{c(b)}-m_{c(b)}^2}{2rr'}-\frac{1}{2r}-\frac{1}{r'}\right],\\
\Pi^{D(B)}_{nonpert2} &= -\langle\bar{s}s\rangle\left[\frac{m_{c(b)}}{rr'}-\frac{m_0^2m_{c(b)}}{24rr'^2}\right],
\end{aligned} \tag{2.4}$$

Here $r=p^2-m_{b(c)}^2$ and $r'=p'^2-m_u^2$. In the above equations, $\rho_1\left(\Pi^{D(B)}_{nonpert1}\right)$ and $\rho_2\left(\Pi^{D(B)}_{nonpert2}\right)$ correspond to the currents $j_\nu^{K_1}$ and $j_{\nu\nu'}^{K_1}$, respectively. The explicit expressions for $I_0(s,s',q^2)$, A, B, C, G and H are presented in [2]. In order to calculate the physical side of the correlation function, for off-shell $D(B)$ case, complete sets of intermediate states are inserted to the aforementioned correlation function. As a result, we get:

$$\begin{aligned}
\Pi^{D(B)}_{\mu\nu} &= \frac{\langle 0|j^{D(B)}|D(B)(q)\rangle\langle D_s^*(B_s^*)(p,\varepsilon)|j_\mu^{D_s^*(B_s^*)}|0\rangle}{(q^2-m_{D(B)}^2)(p^2-m_{D_s^*(B_s^*)}^2)}\left[\frac{\langle 0|j_\nu^{K_1}|K_1^a(p',\varepsilon')\rangle\langle K_1^a(p',\varepsilon')D(B)(q)|D_s^*(B_s^*)(p,\varepsilon)\rangle}{(p'^2-m_{K_1^a}^2)}\right.\\
&\quad\left.+\frac{\langle 0|j_\nu^{K_1}|K_1^b(p',\varepsilon')\rangle\langle K_1^b(p',\varepsilon')D(B)(q)|D_s^*(B_s^*)(p,\varepsilon)\rangle}{(p'^2-m_{K_1^b}^2)}\right]+...,\\
\Pi^{D(B)}_{\mu\nu\nu'} &= \frac{\langle 0|j^{D(B)}|D(B)(q)\rangle\langle D_s^*(B_s^*)(p,\varepsilon)|j_\mu^{D_s^*(B_s^*)}|0\rangle}{(q^2-m_{D(B)}^2)(p^2-m_{D_s^*(B_s^*)}^2)}\left[\frac{\langle 0|j_{\nu\nu'}^{K_1}|K_1^a(p',\varepsilon')\rangle\langle K_1^a(p',\varepsilon')D(B)(q)|D_s^*(B_s^*)(p,\varepsilon)\rangle}{(p'^2-m_{K_1^a}^2)}\right.\\
&\quad\left.+\frac{\langle 0|j_{\nu\nu'}^{K_1}|K_1^b(p',\varepsilon')\rangle\langle K_1^b(p',\varepsilon')D(B)(q)|D_s^*(B_s^*)(p,\varepsilon)\rangle}{(p'^2-m_{K_1^b}^2)}\right]+...,
\end{aligned} \tag{2.5}$$

Here ε and ε' are the polarization vectors associated with D_s^* and $K_1^{a(b)}$ states, respectively. Here, the K_1^a and K_1^b are basic states that we construct the physical $K_1(1270)$ and $K_1(1400)$ states in terms of these states. We also consider the mixture between the $K_1(1270)$ and $K_1(1400)$ axial mesons (for details see [2]). After defining the above matrix elements presented in [2] and summing over the polarization vectors, the final physical representation of the correlation functions are obtain as:

$$\begin{aligned}
\Pi^{D(B)}_{\mu\nu} &= \left(g^{D(B)}_{D_s^*DK_1^a(B_s^*BK_1^a)}(q^2)\frac{m_{K_1^a}f_{K_1^a}}{(p'^2-m_{K_1^a}^2)}+g^{D(B)}_{D_s^*DK_1^b(B_s^*BK_1^b)}(q^2)\frac{m_{K_1^b}f_{K_1^b\perp}a_0^{\parallel,K_1^b}}{(p'^2-m_{K_1^b}^2)}\right)\\
&\quad\times\frac{\frac{f_{D(B)}m_{D(B)}^2}{m_{c(b)}+m_u}f_{D_s^*(B_s^*)}m_{D_s^*(B_s^*)}}{(q^2-m_{D(B)}^2)(p^2-m_{D_s^*(B_s^*)}^2)}\left\{m_{D_s^*(B_s^*)}^2g_{\mu\nu}+\text{other structures}\right\}+...,
\end{aligned}$$

$$\Pi_{\mu\nu\nu'}^{D(B)} = \left(g_{D_s^* D K_1^a (B_s^* B K_1^a)}^{D(B)}(q^2) \frac{f_{K_1^a} a_0^{\perp, K_1^a}}{\left(p'^2 - m_{K_1^a}^2\right)} + g_{D_s^* D K_1^b (B_s^* B K_1^b)}^{D(B)}(q^2) \frac{f_{K_1^{b\perp}}}{\left(p'^2 - m_{K_1^b}^2\right)} \right)$$

$$\times \frac{\frac{f_{D(B)} m_{D(B)}^2}{m_{c(b)} + m_u} f_{D_s^*(B_s^*)} m_{D_s^*(B_s^*)}}{\left(q^2 - m_{D(B)}^2\right)\left(p^2 - m_{D_s^*(B_s^*)}^2\right)} \left\{ m_{D_s^*(B_s^*)}^2 (g_{\mu\nu} p'_{\nu'} - g_{\mu\nu'} p'_{\nu} + \text{other structures} \right\} + ... \quad (2.6)$$

where the $f_{K_1^a}$, $f_{K_1^{b\perp}}$, $f_{D(B)}$ and $f_{D_s^*(B_s^*)}$ are decay constants, The $a_0^{\perp, K_1^a}$ and $a_0^{\|, K_1^b}$ are zeroth order Gegenbauer moments. In order to compute the coupling constants, we choose the structures $g_{\mu\nu}$ and $g_{\mu\nu} p'_{\nu'}$ from both sides of the correlation function. Equating the physical and QCD sides of the correlation function we get the QCD sum rules for the considered strong coupling constants form factors. To suppress contributions of the higher states and continuum, we apply double Borel transformation as well as continuum subtraction. These processes bring two Borel parameters and two continuum thresholds which we find their working regions [2]. Similarly, one can get the sum rules for coupling form factors when the K_1 meson is as the off-shell state.

3. Numerical Analysis

Having determined the QCD sum rules for the coupling constants, it is time to look at how the coupling constants behave in terms of Q^2 where $Q^2 = -q^2$ (see [2] for details). Using the working region of the auxiliary parameters as well as other input parameters, we find the coupling form factors in terms of Q^2 as presented in [2]. The coupling constants are the values of the coupling form factors at $Q^2 = -m_{off-shell}^2$. Finally, we obtain the results as presented in Tables 1 and 2. The final result for each vertex is obtained taking the average of the two corresponding off-shell states.

Table 1. Values of the $g_{D_s^* D K_1(1270)}$ and $g_{D_s^* D K_1(1400)}$ coupling constants in GeV^{-1} unit.

	$Q^2 = -m_D^2$	$Q^2 = -m_{K_1(1270)}^2$	Average
$g_{D_s^* D K_1(1270)}$	2.83 ± 0.09	1.36 ± 0.14	2.09 ± 0.82
	$Q^2 = -m_D^2$	$Q^2 = -m_{K_1(1400)}^2$	Average
$g_{D_s^* D K_1(1400)}$	0.97 ± 0.15	1.12 ± 0.54	1.04 ± 0.78

Table 2. Values of the $g_{B_s^* B K_1(1270)}$ and $g_{B_s^* B K_1(1400)}$ coupling constants in GeV^{-1} unit.

	$Q^2 = -m_B^2$	$Q^2 = -m_{K_1(1270)}^2$	Average
$g_{B_s^* B K_1(1270)}$	1.18 ± 0.07	0.81 ± 0.45	1.99 ± 0.11
	$Q^2 = -m_B^2$	$Q^2 = -m_{K_1(1400)}^2$	Average
$g_{B_s^* B K_1(1400)}$	0.35 ± 0.05	0.26 ± 0.04	0.30 ± 0.05

4. Acknowledgement

This work has been supported partly by the Scientific and Technological Research Council of Turkey (TUBITAK) under the research project 110T284.

References

[1] Shifman M A, Vainshtein A I and Zakharov V I, *Nucl. Phys. B* **147**, 385-447, 1979.
[2] Sundu H, Sungu J Y, Sahin S, Yinelek N and Azizi K, *Phys. Rev. D* **83**:114009, 2011.

Journal of Physics: Conference Series **348** (2012) 012011 doi:10.1088/1742-6596/348/1/012011

$g_{D_sDK_0^*}$ and $g_{B_sBK_0^*}$ coupling constants in QCD sum rules

S Şahin*[1], H Sundu[1] and K Azizi[2]

[1]Department of Physics, Kocaeli University, 41380 Izmit, Turkey
[2]Department of Physics, Doğuş University, Acıbadem-Kadıköy, 34722 Istanbul, Turkey

E-mail: *[1]physics.sinemsahin@gmail.com

Abstract. In the present study, we calculate the strong coupling constants $g_{D_sDK_0^*(800)}$ and $g_{B_sBK_0^*(800)}$ within the three-point QCD sum rules approach. We evaluate the correlation function of the considered vertices taking into account both D[B] and $K_0^*(800)$ mesons as off-shell states.

1. Introduction

In low energies, it is difficult to obtain confident theoretical results using the perturbation theory since the interaction between quarks and gluons becomes large in this scale. Therefore, we need some non-perturbative approaches to describe low energy dynamics of hadrons clearly. Among these dynamics, the strong coupling constants of mesons are closely related to their strong interactions. Using the QCD sum rules method as one of the powerful and applicable non-perturbative approaches [1], we can determine these coupling constants more accurately. In this work, we calculate the strong coupling constants of $D_sDK_0^*(800)$ and $B_sBK_0^*(800)$ vertices. Calculation of such coupling constants can help us in understanding the nature of the strong interaction among the participating particles.

2. QCD Sum Rules for the Strong Coupling Constants

In this section, we obtain QCD sum rules for the strong coupling constants associated with the $D_s - D - K_0^*(800)$ and $B_s - B - K_0^*(800)$ vertices. For this aim, the following three-point correlation function for $D(B)$ off-shell case is studied:

$$\Pi^{D(B)} = i^2 \int d^4x \, d^4y \, e^{ip'\cdot x} \, e^{iq\cdot y} \langle 0| \mathcal{T} \left(\eta^{K_0^*}(x) \, \eta^{D(B)}(y) \, \eta^{D_s(B_s)\dagger}(0) \right) |0\rangle, \tag{1}$$

where $\mathcal{T}$ indicates the time ordering product, $q = p - p'$ is the momentum of the off-shell state and p' is the momentum of the final on-shell state. The interpolating quark currents, which produce the considered mesons from the vacuum with the same quantum numbers as these currents can be written in terms of the quark field operators as shown in [2]. In the QCD sum rules method, we calculate the related correlation function in two different ways as outlined in [2]. The strong coupling constants are obtained equating these two different presentations via dispersion relation. To suppress contributions of the higher states and continuum, we will apply double Borel transformation with respect to the momentum squared of the initial and final on-shell states to both sides of the obtained sum rules.

First, we start to the calculation of the physical side of the concerned correlation function. Inserting a complete set of intermediate hadronic states with the same quantum numbers as interpolating quark currents into the correlation function, we obtain:

$$\Pi^{D(B)} = \frac{\langle 0|\eta^{K_0^*}|K_0^*(p')\rangle\langle 0|\eta^{D(B)}|D(B)(q)\rangle\langle K_0^*(p')D(B)(q)|D_s(B_s)(p)\rangle\langle D_s(B_s)(p)|\eta^{D_s(B_s)}|0\rangle}{(q^2-m_{D(B)}^2)(p^2-m_{D_s(B_s)}^2)(p'^2-m_{K_0^*}^2)} + \ldots, \quad (2)$$

The matrix elements shown in the above equation can be parameterized in terms of leptonic decay constants and strong coupling constant $g^{D(B)}_{D_sDK_0^*(B_sBK_0^*)}(q^2)$ (see [2] for details). The final physical representation of the correlation function in the case of D(B) off-shell is obtained as:

$$\Pi^{D(B)} = g^{D(B)}_{D_sDK_0^*(B_sBK_0^*)}(q^2)\frac{f_{K_0^*}m_{K_0^*}\frac{f_{D(B)}m_{D(B)}^2}{m_{c(b)}+m_u}\frac{f_{D_s(B_s)}m_{D_s(B_s)}^2}{m_{c(b)}+m_s}}{2\,(q^2-m_{D(B)}^2)(p'^2-m_{K_0^*}^2)(p^2-m_{D_s(B_s)}^2)}(m_{D_s(B_s)}^2+m_{K_0^*}^2-q^2) + \ldots., \quad (3)$$

Now, we concentrate to calculate the QCD or theoretical side of the considered correlation function. The correlation function in QCD side is written in terms of the perturbative and non-perturbative parts. The perturbative part is defined in terms of double dispersion integral as shown in [2]. In order to obtain the spectral density, we need to calculate the bare loop diagram for $D(B)$ off-shell presented in [2]. As a result, the spectral density is obtained as follows:

$$\rho^{D(B)}(s,s',q^2) = \frac{N_c}{2\,\lambda^{1/2}(s,s',q^2)}\Big\{m_s\Big(m_u(m_s+m_u)-q^2\Big)-sm_u - m_{c(b)}\Big((m_s+m_u)^2-s'-m_{c(b)}(m_s+m_u)\Big)\Big\}, \quad (4)$$

Now, we proceed to calculate the nonperturbative contributions in QCD side. We consider the quark-quark and quark-gluon condensate diagrams presented in [2]. As a result, we obtain:

$$\Pi^{D(B)}_{nonper} = \frac{\langle\bar{s}s\rangle}{2}\Big\{\frac{2m_{c(b)}m_u-m_{c(b)}^2+q^2}{rr'}-\frac{1}{r}-\frac{1}{r'}+\frac{m_0^2(4m_{c(b)}m_u-m_{c(b)}^2+q^2)}{4r^2r'} - \frac{m_0^2}{4rr'}-\frac{m_0^2}{4r^2}+\frac{m_0^2(m_{c(b)}^2-4m_{c(b)}m_u-q^2)}{4rr'^2}+\frac{m_0^2}{4r^2}+\frac{m_0^2}{4rr'}\Big\}, \quad (5)$$

Finally, after applying the double Borel transformation, the following sum rules for the considered coupling constant is obtained as:

$$g^{D(B)}_{D_sDK_0^*(B_sBK_0^*)}(q^2) = \frac{2(q^2-m_{D(B)}^2)(m_{c(b)}+m_u)(m_{c(b)}+m_s)}{m_{D_s(B_s)}^2m_{D(B)}^2m_{K_0^*}f_{D_s(B_s)}f_{D(B)}f_{K_0^*}(m_{D_s(B_s)}^2+m_{K_0^*}^2-q^2)}e^{\frac{m_{D_s(B_s)}^2}{M^2}} \times e^{\frac{m_{K_0^*}^2}{M'^2}}\Big[-\frac{1}{4\pi^2}\int_{(m_{c(b)}+m_s)^2}^{s_0}ds\int_{(m_s+m_u)^2}^{s'_0}ds'\rho^{D(B)}(s,s',q^2)\theta[1-(f^{D(B)}(s,s'))^2]e^{\frac{-s}{M^2}}e^{\frac{-s'}{M'^2}} + \widehat{B}\Pi^{D(B)}_{nonper}\Big], \quad (6)$$

Similarly, one can calculate the coupling constant for $K_0^*(800)$ off-shell case.

3. Numerical analysis

To obtain the behavior of the coupling constants in terms of q^2, we should look for working regions for Borel mass parameters (M^2 and M'^2)and continuum thresholds (s_0 and s'_0) (see [2] for details). We use the following Boltzmann function to find the Q^2 behavior of the considered strong coupling constants:

$$g(Q^2) = \left[A_1 + \frac{A_2}{1+\exp[\frac{Q^2-x_0}{\Delta x}]}\right] [GeV^{-1}]. \tag{7}$$

where, $Q^2 = -q^2$ and the parameters A_1, A_2, x_0, Δx are given in Table 1.

Table 1. Parameters appearing in the fit function of the coupling constants for $D_sDK_0^*(800)$ and $B_sBK_0^*(800)$ vertices.

	A_1	A_1	x_0	Δx
$g^{(D)}_{D_sDK_0^*(800)}(Q^2)$	3.468	-2.741	8.067	4.995
$g^{(K_0^*(800))}_{D_sDK_0^*(800)}(Q^2)$	-0.024	0.772	5.723	1.257
$g^{(B)}_{B_sBK_0^*(800)}(Q^2)$	4.151	-1.932	13.842	12.149
$g^{(K_0^*(800))}_{B_sBK_0^*(800)}(Q^2)$	-0.017	0.547	5.431	1.121

Table 2. Value of the $g_{D_sDK_0^*(800)}$ and $g_{B_sBK_0^*(800)}$ coupling constants in GeV^{-1} unit.

	$Q^2 = -m_D^2$	$Q^2 = -m^2_{K_0^*(800)}$	Average
$g_{D_sDK_0^*(800)}$	0.97 ± 0.02	0.74 ± 0.05	0.85 ± 0.08
	$Q^2 = -m_B^2$	$Q^2 = -m^2_{K_0^*(800)}$	Average
$g_{B_sBK_0^*(800)}$	2.28 ± 0.18	0.53 ± 0.09	1.41 ± 0.21

The final result for each coupling constant is obtained taking the average of the coupling constants obtained from two different off-shell cases.

4. Acknowledgement

This work is supported in part by the Scientific and Technological Research Council of Turkey (TUBITAK) under the research project No. 110T284 and Research Fund of Kocaeli University under grant no. 2011/029.

References

[1] Shifman M A, Vainstein A I and Zakharov V I 1979 *Nucl. Phys. B* **147** 385 ; Shifman M A, Vainstein A I and Zakharov V I 1979 *Nucl. Phys. B* **147** 448
[2] Sundu H, Sungu J Y, Sahin S, Yinelek N, Azizi K 2011 *Phys.Rev.D* **26**

Theoretical determination of $K_1(1270, 1400)$ mixing angle in QCD

Hüseyin Dağ[1,2,*], Altuğ Özpineci[3], Ayçe Çağıl[3], Güray Erkol[2]

[1] Middle East Technical University, North Cyprus Campus, Kalkanlı, Güzelyurt, KKTC, Mersin 10, Turkey.
[2] Laboratory for Fundamental Research , Ozyegin University, Nisantepe Mah. Orman Sok. No:13, 34794 Alemdag-Cekmekoy, Istanbul, Turkey.
[3] Department of Physics, Middle East Technical University, 06531 Ankara, Turkey.
[*] Presenter

E-mail: hdag@metu.edu.tr, ozpineci@metu.edu.tr, ayse.cagil@cern.ch, guray.erkol@ozyegin.edu.tr

Abstract. In quark model, the strange axial vector mesons $K_1(1270)$ and $K_1(1400)$ are defined as the mixtures of orbital angular momentum states K_{1A} and K_{1B}. In this work, by using the orthogonality of the mass eigenstates, we have estimated the $K_1(1270, 1400)$ mixing angle θ_{K_1}, where we have found that $\theta_{K_1} \simeq -(39 \pm 4)^\circ$.

The Particle Data Group lists the axial vector, $I = 1/2$, P-wave strange mesons $K_1(1270)$ and $K_1(1400)$ with masses 1273 ± 7 MeV and 1402 ± 7 MeV, respectively [1]. The decays of these mesons are constrained by a selection rule such that $\Gamma(K_1(1270) \to K\rho) \gg \Gamma(K_1(1270) \to K^*\pi)$ and $\Gamma(K_1(1400) \to K^*\pi) \gg \Gamma(K_1(1400) \to K\rho)$. This rule suggests a large mixing angle close to $|45|^\circ$ between the pure orbital angular momentum and the G-parity states, $K_{1A}(1^3P_1)$ and $K_{1B}(1^1P_1)$, leading to the physical $K_1(1270)$ and $K_1(1400)$ states [2, 3]. This mixing is written as

$$\begin{aligned} |K_1(1270)\rangle &= \sin\theta_{K_1}|K_{1A}\rangle + \cos\theta_{K_1}|K_{1B}\rangle, \\ |K_1(1400)\rangle &= \cos\theta_{K_1}|K_{1A}\rangle - \sin\theta_{K_1}|K_{1B}\rangle, \end{aligned} \tag{1}$$

where θ_{K_1} is the mixing angle [4]. An accurate determination of θ_{K_1} is crucial to understand the physical properties of the axial-vector strange mesons. There are various approaches to this problem in the literature. In 1977, Carnegie *et al.* obtained $\theta_{K_1} = (41 \pm 4)^\circ$ by making a fit to the available data [3]. Brundell *et al.* estimated the mixing angle to be in the range $-30^\circ < \theta_{K_1} < 50^\circ$ by using the data on the ratio $BR(\tau \to K_1(1270)\nu_\tau)/BR(\tau \to K_1(1400)\nu_\tau)$. The favored value in this range from their analysis is $\theta_{K_1} \simeq 45^\circ$ [5]. It was found from experimental data that the mixing angle has a value $|\theta_{K_1}| \simeq 32^{+8^\circ}_{-2}$ or $|\theta_{K_1}| \simeq 57^{+2^\circ}_{-3}$ and among these $|\theta_{K_1}| \simeq 33^\circ$ is favored [6]. Burakovsky *et al* used a nonrelativistic constituent quark model and estimated $35^\circ < |\theta_{K_1}| < 55^\circ$[7]. In addition to these analyses the mixing angle has been extracted from the updated data on $\tau \to K_1(1270)\nu_\tau$ decays as $|\theta_{K_1}| \simeq 37^\circ$ or $|\theta_{K_1}| \simeq 58^\circ$[8].

It is apparent from above results in the literature that there exists a sign ambiguity in θ_{K_1} due to different definitions. In Ref. [9], the decay constants of K_{1A} and K_{1B} are defined positive and the sign ambiguity is removed which results in a negative value of θ_{K_1}. Hatanaka and Yang analyzed the latest data on $B \to K_1(1270,1400)\gamma$ and $\tau \to K_1(1270,1400)\nu_\tau$ and they have found $\theta_{K_1} = -(34 \pm 13)^\circ$.

In the present work we have used a theoretical approach, which has been introduced by T. M. Aliev, A. Ozpineci and V.S. Zamiralov [10], to determine the mixing angle θ_{K_1}. In this approach, the correlator function is constructed in terms of the interpolating currents of the pure K_{1A} and K_{1B} states. Then the mixing angle can be calculated using the orthogonality of the mass eigenstates, *i.e.*, $K_1(1270)$ and $K_1(1400)$. Using operator product expansion, the mixing is expressed in terms of quark and gluon degrees of freedom.

We start our analysis with the correlator function

$$\begin{aligned} \Pi_{\mu\nu}(p^2) &\equiv \Pi_{\mu\nu}^{K_1(1270)K_1(1400)}(p^2) \\ &= i\int d^4x e^{ipx}\langle 0|\mathcal{T}\{j_\mu^{K_1(1400)\dagger}(x) j_\nu^{K_1(1270)}(0)\}|0\rangle \;=\; 0+\ldots \end{aligned} \tag{2}$$

which should vanish since there is no transition possible between the mass eigenstates. To calculate the correlator in Eq. (2), the interpolating currents of $K_1(1270,1400)$ mesons should be written in terms of pure states $K_{1(A,B)}$. Using the definition of mixing in Eq. (1), we obtain the relation

$$\begin{aligned} j^{K_1(1270)} &= \sin\theta_{K_1} j^{K_{1A}} + \cos j^{K_{1B}}, \\ j^{K_1(1400)} &= \cos\theta_{K_1} j^{K_{1A}} - \sin j^{K_{1B}}, \end{aligned} \tag{3}$$

where $j^{K_{1(A,B)}}$ are the currents for the pure states $K_{1(A,B)}$. In the SU(3) limit, K_{1A} and K_{1B} couple to axial-vector and tensor currents respectively. When SU(3) symmetry is broken, the contributions to pure currents are proportional to the Gagenbauer moments, and since they are either zero or comparable to zero, we can assume the pure axial-vector and tensor currents for K_{1A} and K_{1B}, respectively [4, 11]. Then we write

$$|K_{1A}(p,\epsilon)\rangle = \lambda_A \eta_\mu^A|0\rangle \quad \text{and} \quad |K_{1B}(p,\epsilon)\rangle = \lambda_B \eta_\mu^B|0\rangle, \tag{4}$$

where $\lambda_A = (if_A m_A)^{-1}$, $\lambda_B = (f_B m_B^2)^{-1}$ are the coefficients, $\eta_\mu^A = \bar{s}\gamma_\mu\gamma_5 d$ and $\eta_\mu^B = \bar{s}\sigma_{\mu\alpha}p^\alpha\gamma_5 d$ are the pure axial-vector and tensor currents. Here $f_{A,B}$ and $m_{A,B}$ are the decay constants and masses of $K_{1(A,B)}$ states, respectively. After inserting the definitions of mixing in Eq. (3) and the pure currents in Eq. (4), the correlator in Eq. (2) can be written in terms of the mixing angle as follows:

$$\tan\theta_{K_1}(\lambda_A^2\Pi_{\mu\nu}^{AA} - \lambda_B^2\Pi_{\mu\nu}^{BB}) + (1+\tan^2\theta_{K_1})\lambda_A\lambda_B\Pi_{\mu\nu}^{AB} = 0 + \ldots\,, \tag{5}$$

where

$$\Pi_{\mu\nu}^{ij}(p^2) = i\int dx e^{ipx}\ \mathrm{Tr}[\Upsilon_\mu^{i\dagger}(x)(iS_d(-x))\Upsilon_\nu^j(0)(iS_s(x))] \tag{6}$$

is the two-point correlator of pure states with $i,j = A,B$, $\Upsilon_\mu^A = \gamma_\mu\gamma_5$ and $\Upsilon_\mu^B = \sigma_{\mu\alpha}p^\alpha\gamma_5$ are the Lorentz structures in pure currents and $iS_q(x)$ is the full quark propagator in position space. The correlator for pure states can be expressed in terms of Lorentz coefficients as $\Pi_{\mu\nu}^{ij} = \Gamma^{ij\prime}g_{\mu\nu} + \Gamma^{ij}p_\mu p_\nu$. We choose the $p_\mu p_\nu$ structure and apply a Borel transformation with respect to p^2. Then Eq. (5) results in an expression for the mixing angle as

$$\tan(2\theta_{K_1}) = \frac{-2\lambda_A^\dagger\lambda_B\bar{\Gamma}^{AB}}{\lambda_A^2\bar{\Gamma}^{AA} - \lambda_B^2\bar{\Gamma}^{BB}}, \tag{7}$$

where $\bar{\Gamma}$ is the Borel transformed Γ with respect to p^2. Here M^2 is the Borel mass and s_0 is the continuum threshold.

In this work we have taken the following numerical inputs:

$$m_s = 100^{+30}_{-20}\ \text{MeV}, f_A = 250 \pm 13\ \text{MeV},\ f_B = 190 \pm 10\ \text{MeV},\ m_A = 1.31\text{GeV}, m_B = 1.34\ \text{GeV}. \quad (8)$$

Our numerical results for $\tan 2\theta_{K_1}$ and θ_{K_1} are shown in figures 1(a) and 1(b). We plot the dependence of $\tan 2\theta_{K_1}$ and θ_{K_1} on M^2 for different values of s_0. The plots suggest that $\tan 2\theta_{K_1}$ lies in the region $-13 \leq \tan 2\theta_{K_1} \leq -3$ and $\theta_{K_1} = -(39 \pm 4)^\circ$. This value is consistent with the previous results in the literature. We will give our detailed analytical expressions and improved analysis in a future work.

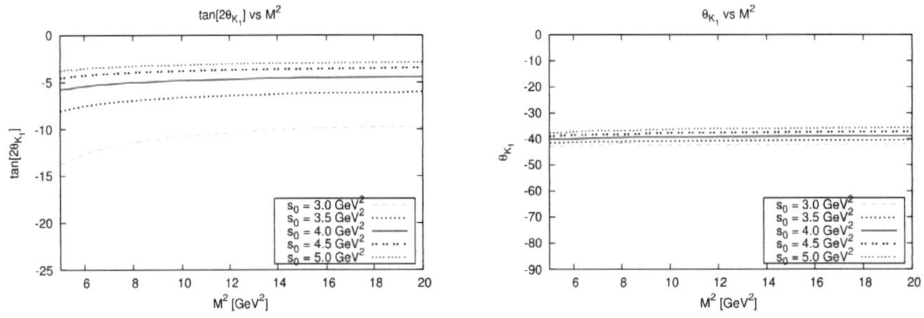

(a) (b)

Figure 1. (a)$\tan 2\theta_{K_1}$ vs. M^2 and (b) θ_{K_1} vs M^2, for different values of s_0.

References

[1] K. Nakamura et al. (Particle Data Group), J. Phys. G **37**, 075021 (2010).
[2] G. W. Brandenburg et al., Phys. Rev Lett. **36** (1976) 703; M.G. Bowler, J. Phys. G **3** (1977) 775.
[3] R.K. Carnegie et al., Nucl. Phys. B **127** (1977) 509.
[4] K-C. Yang, Nucl. Phys. B **776** (2007) 187, `arXiv/hep-ph:0705.0692`.
[5] H.G. Brundell et al., Phys. Rev. D **53** (1996) 3712, `arXiv/hep-ph:9510245`.
[6] M. Suzuki et al., Phys. Rev. D **47** (1993) 1252.
[7] L. Burakovsky and T. Goldman, Phys. Rev. D **57** (1998) 2879; L. Burakovsky and T. Goldman, Phys. Rev. D **56** (1997) 1368, `arXiv/hep-ph:9703274`
[8] H. Y. Cheng, Phys. Rev. D **67** (2003) 094007.
[9] H. Hatanaka, K.-C. Yang, Phys. Rev. D **77** (2008) 094023.
[10] T. M. Aliev, A. Ozpineci and V. Zamiralov, Phys.Rev. D**83** (2011) 016008, `arXiv/hep-ph:1007.0814`.
[11] H. Dag, A. Ozpineci and M.T. Zeyrek, J.Phys.G G**38** (2011) 015002, `arXiv/hep-ph:1001.0939`.

Hadronic Interactions from SU(2) Lattice QCD

Toru T. Takahashi

Gunma National College of Technology, Toribamachi 580, Maebashi, Gunma 371-8530, Japan

Yoshiko Kanada-En'yo

Department of Physics, Kyoto University, Sakyo, Kyoto 606-8502, Japan

Abstract. We study hadronic interactions from Bethe-Salpeter amplitudes in quenched two-color lattice QCD, concentrating on the interactions of two scalar diquarks ($uC\gamma_5 d$) in S-wave scattering states. Between two identical scalar diquarks, we observe repulsive force in short-range region. By defining and evaluating the "quark-exchange part" in the interaction, which is induced by quark-exchange diagrams, or equivalently, by introducing Pauli blocking among some of quarks, we find that the repulsive force in short-distance region arises purely from the "quark-exchange part" and that it disappears when quark-exchange diagrams are omitted. It is qualitatively consistent with the constituent-quark model picture that the origin of short-range repulsion is a color-magnetic interaction among quarks. We also find a novel long-ranged attractive force, which enters in any flavor channels of two scalar diquarks and whose interaction range and strength are quark-mass independent.

1. Introduction

Hadronic interactions are of great importance in nuclear and hadron physics, and baryon-baryon interaction is one of the most significant issues to be clarified in reliable nonperturbative manners. After the successful proposal by Yukawa, the long- and the intermediate-range interactions have been well described by one-boson-exchange potentials (OBEP), in which key players for the hadronic interactions are mesonic modes. On the other hand, the short-distance part of hadronic interactions has not been well clarified so far. Especially, the origin of the strong repulsive core in nucleon-nucleon (NN) channel [2], which is responsible for the high-density nuclear phenomena, has been one of the longstanding problems in nuclear and hadron physics [3, 4, 5, 6, 7, 8, 9, 10].

Whereas it has been long since Quantum ChromoDynamics (QCD), the fundamental theory of strong interactions, was established as the fundamental theory, QCD-based description of hadronic dynamics has not yet been successful. Recently, a nucleon-nucleon (NN) interaction was evaluated from NN Bethe-Salpeter (BS) amplitudes on the lattice [11], and the authors observed the short-ranged repulsive force and the attraction in the intermediate region. In order to clarify the essential structure of hadronic interactions, we study hadronic potential using SU(2) lattice QCD [1], which is simpler than SU(3) QCD and can be a testbed for understanding hadron interactions.

2. Simulation Setup

We follow the strategy proposed by CP-PACS group [1, 12], where asymptotic Bethe-Salpeter wavefunctions on the Euclidean lattice are adopted to evaluate pion scattering length. This method has been further developed and employed for various hadron-hadron channels or interquark potentials [13, 14, 15].

We measure Bethe-Salpeter wavefunctions between two scalar-diquarks ($qC\Gamma_5 q$), and define an interhadron "potential" assuming a nonrelativistic Schrödinger-type equation. We focus on S-wave scattering states of two scalar diquarks by projecting BS-wavefunctions onto A_1^+-wavefunctions, which have overlap with $l = 0$ states. Simulations are performed in SU(2) quenched QCD adopting the standard plaquette gauge action and the Wilson quark action. The lattice extent is $24^3 \times 64$ at $\beta = 2.45$ and the corresponding physical lattice spacing is about 0.1 fm, which is determined to reproduce the string tension $\sqrt{\sigma}$ of 440 MeV. We use four different hopping parameters (quark masses) κ = 0.1350, 0.1400, 0.1450, 0.1500.

We investigate two-scalar-diquark scattering between $q_i C\Gamma_5 q_j$ and $q_k C\Gamma_5 q_l$, where i, j, k, l denote the flavors of quarks. We consider two different flavor combinations. One combination is $(i, j, k, l) = (1, 2, 1, 2)$, where only two independent flavors exist in four quarks. The other is $(i, j, k, l) = (1, 2, 3, 4)$, where all the quarks have different flavors. We adopt the same hopping parameters (quark masses) for all the quarks in both cases, and therefore all the scalar diquarks degenerate in mass. "Quark-exchange diagrams" in quark-propagator contraction are included only in the $(i, j, k, l) = (1, 2, 1, 2)$ case. The existence of quark-exchange diagrams are considered to be essential for the short-range interactions, since any Pauli-blocking effects among quarks are not included without exchange diagrams. We expect that the origin of short-range hadronic interactions can be accessed by comparing these two flavor combinations.

3. Lattice QCD results

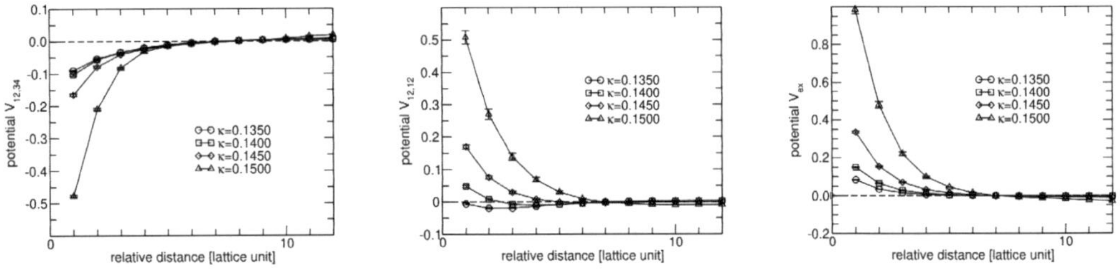

Figure 1. *Left*: Potentials $V_{12,34}(R)$ computed with the flavor combination, $(i, j, k, l) = (1, 2, 3, 4)$, are plotted as functions of relative distance R. *Middle*: Potentials $V_{12,12}(R)$ computed with the flavor combination, $(i, j, k, l) = (1, 2, 1, 2)$, are plotted as functions of relative distance R. *Right*: "Quark-exchange parts" of potentials, which are defined as $V_{\rm ex}(R) \equiv V_{12,12}(R) - V_{12,34}(R)$, are plotted as functions of relative distance R.

3.1. Potentials : $V_{12,34}(R)$ and $V_{12,12}(R)$

Fig. 1 [left] shows the reconstructed potentials $V_{12,34}(R, m_q)$ as functions of the interhadron distance R, which are extracted with the flavor combination $(i, j, k, l) = (1, 2, 3, 4)$. Here, we include no quark-exchange diagram and hence no Pauli-blocking effect among quarks is activated. One finds that the interaction in this channel is always attractive, and that the strength of this attraction depends on quark masses. $V_{12,34}(R, m_q)$ rapidly reduces with decreasing κ (increasing m_q), and finally the potentials at κ=0.1350 and 0.1400 coincide with each other, which implies the existence of a long-range *quark-mass independent* attraction.

The interhadron potentials $V_{12,12}(R, m_q)$ obtained with $(i, j, k, l) = (1, 2, 1, 2)$ can be seen in Fig. 1 [middle]. The important difference from the previous $(i, j, k, l) = (1, 2, 3, 4)$ case is quark-exchange diagrams newly included, which gives rise to Pauli-blocking effect among quarks. We now find *strong repulsions in the short-distance region.*

3.2. Potentials : Quark-exchange part $V_{\rm ex}(R)$

In the following, we define a "quark-exchange part" $V_{\rm ex}(R, m_q)$ in $V_{12,12}(R, m_q)$ as

$$V_{\rm ex}(R, m_q) \equiv V_{12,12}(R, m_q) - V_{12,34}(R, m_q), \tag{1}$$

which is nothing but the difference between $V_{12,12}(R, m_q)$ and $V_{12,34}(R, m_q)$. Taking into account that quark-exchange diagrams are included only in $V_{12,12}(R, m_q)$ and that the diagrams needed for $V_{12,34}(R, m_q)$ are identical with the direct diagrams in $V_{12,12}(R, m_q)$, it can measure "quark-exchange effect" or Pauli-blocking effect among quarks in $V_{12,12}(R, m_q)$. In practice, it is equivalent to assuming

$$V_{12,34}(R, m_q) = V_{\rm dir}(R, m_q), \quad V_{12,12}(R, m_q) = V_{\rm dir}(R, m_q) + V_{\rm ex}(R, m_q), \tag{2}$$

with a "direct part" $V_{\rm dir}(R, m_q)$ measured only with direct diagrams and a "quark-exchange part" $V_{\rm ex}(R, m_q)$ induced by adding exchange diagrams. Such decomposition of a diquark-diquark potential is conceptually similar to the usual decomposition procedure in RGM (Resonating Group Method) calculations in quark cluster models [7, 16]. Thus extracted potential $V_{\rm ex}(R, m_q) \equiv V_{12,12}(R, m_q) - V_{12,34}(R, m_q)$ shows a monotonous behavior as is seen in Fig. 1 [right], and therefore the decomposition seems beneficial for our purpose to clarify strengths or ranges of potentials. One can now find that *the short-range repulsion arises only from* $V_{\rm ex}(R, m_q)$, which implies Pauli-blocking effect among quarks is essential for the short-range repulsion.

4. Discussions

We have found that $V_{12,12}(R, m_q)$ and $V_{12,34}(R, m_q)$ can be expressed as shown in Eq.(2). The key property of the R-dependent potentials is twofold; *interaction ranges* and *strengths*. Hadronic interactions are phenomenologically incorporated as one-boson-exchange-potentials (OBEP), which are Yukawa-type and whose interaction ranges are directly connected to exchanged meson masses. Then, such OBE potentials should be sensitive to meson masses, *i.e.* quark masses. On the other hand, the short-range repulsive core is often considered to arise from a color-magnetic interaction induced by one-gluon-exchanges (OGE), whose strength is proportional to the inverse square of (constituent) quark masses, $\sim m_{\rm Q}^{-2}$. Such $m_{\rm Q}$-dependent interactions could be clarified by monitoring the strengths of the potentials.

We evaluate strengths and ranges of these decomposed potentials, $V_{\rm dir}(R, m_q)$ and $V_{\rm ex}(R, m_q)$, by fitting them with an exponential form, $F(x) \equiv A\exp\left(-\frac{x}{B}\right)$, which have two parameters, strength A and range B. (We also tried several other functions. See Ref. [1].) The fit results can be all found in Fig. 2, in which strengths A and ranges B for $V_{\rm dir}(R, m_q)$ and $V_{\rm ex}(R, m_q)$ at each κ are displayed.

4.1. strengths and ranges : $V_{\rm dir}$

In Fig. 2, the fitted parameters, strength $A_{\rm dir}$ and range $B_{\rm dir}$ of the attractive part $V_{\rm dir}(R, m_q)$, are shown as functions of a constituent quark mass $m_{\rm Q}$, which is defined by a half of axialvector-diquark mass in this paper. (Shown as "DIR" in Fig. 2.) Interestingly, both of the strength and the range of $V_{\rm dir}(R, m_q)$ exhibit flattening in the heavy quark-mass region, which indicates that $V_{\rm dir}(R, m_q)$ contains a universal attractive potential $V^U_{\rm att}(R)$. "Universal" here means that neither the strength nor the range of $V^U_{\rm att}(R)$ depends on quark mass and that $V^U_{\rm att}(R)$ always appears in any flavor channels.

We further define a quark-mass dependent part $V^D_{\rm att}(R, m_q)$ as

$$V^D_{\rm att}(R, m_q) \equiv V_{\rm dir}(R, m_q) - V^U_{\rm att}(R). \tag{3}$$

We also fit $V^D_{\rm att}(R, m_q)$ with $F(x)$, and the fitted parameters are shown in Fig. 2. We simply adopt $V_{\rm dir}(R, m_q)$ at $\kappa = 0.1350$ as the universal part $V^U_{\rm att}(R)$, since we observe almost no quark-mass dependence already at this κ. The strength and the range of $V^D_{\rm att}(R, m_q)$ are shown as "ATT(D)". The interaction range $B^D_{\rm att}$ no longer depends on quark mass. That is, $V_{\rm dir}(R, m_q)$ is approximately described by two independent parts;

$$V_{\rm dir}(R, m_q) \sim V^U_{\rm att}(R) + V^D_{\rm att}(R, m_q) = A^U_{\rm att} f^U_{\rm att}(R) + A^D_{\rm att}(m_q) f^D_{\rm att}(R). \tag{4}$$

First part represents a m_q-independent weak long-range force, and the second one does a short-range repulsive interaction that has an m_q-dependent strength and an m_q-insensitive interaction range.

4.2. strengths and ranges : $V_{\rm ex}$

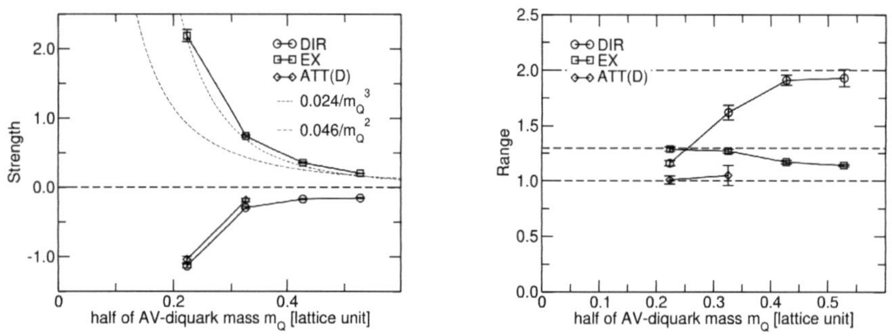

Figure 2. *Left*: The fitted strengths, $A_{\rm dir}$, $A_{\rm ex}$, and $A^D_{\rm att}$, of the potentials, $V_{\rm dir}(R, m_q)$, $V_{\rm ex}(R, m_q)$, and $V^D_{\rm att}(R, m_q)$, are plotted as functions of half of axialvector-diquark mass m_Q. *Right*: The fitted interaction ranges, $B_{\rm dir}$, $B_{\rm ex}$, and $B^D_{\rm att}$, of the potentials, $V_{\rm dir}(R, m_q)$, $V_{\rm ex}(R, m_q)$, and $V^D_{\rm att}(R, m_q)$. The parameters for $V_{\rm dir}(R, m_q)$, $V_{\rm ex}(R, m_q)$, and $V^D_{\rm att}(R, m_q)$ are respectively shown as "DIR", "EX", and "ATT(D)".

The fitted parameters for the quark-exchange part $V_{\rm ex}(R, m_q)$ are shown in Fig. 2. (Shown as "EX" in Fig. 2.) Let us first have a look at the range parameter $B_{\rm ex}$, which are shown as "EX" in Fig. 2 [right]. One can find that the range $B_{\rm ex}$ in $V_{\rm ex}(R, m_q)$ is almost quark-mass independent. The constituent quark-mass (half of AV-diquark mass) variation from 0.53 to 0.22 gives rise to only 10% change in the range $B_{\rm ex}$, which is much smaller than expected from the meson-mass change. Moreover, at two largest κ's (two lightest quark masses), the range $B_{\rm ex}$ remains unchanged. $V_{\rm ex}(R, m_q)$ can be approximately expressed as

$$V_{\rm ex}(R, m_q) \sim A_{\rm ex}(m_q) f_{\rm ex}(R). \tag{5}$$

The quark-mass dependence appears only in the strength $A_{\rm ex}$. From these observations, we conjecture that the short-range repulsion observed here is not generated by meson poles as OBEP, but by some other origins.

One most possible mechanism for the short-range repulsion is a color-magnetic (CM) interaction among quarks as suggested by constituent-quark models. A CM interaction accompanied by Pauli blocking among quarks raises the energy of closely located two hadrons, which results in strong repulsions. In the first-order perturbation, the contributions from the CM interaction are proportional to its strength, whose m_q-dependence is given by m_Q^{-2}.

The strengths $A_{\rm ex}$ are also shown in Fig. 2 [left]. $A_{\rm ex}$ increases in the light-quark-mass region and it is qualitatively consistent with the CM interaction as the origin of repulsion. In Fig. 2 [left], two fit functions, Cm_Q^{-2} and Cm_Q^{-3}, are plotted as a long-dashed and a dashed line. The strength $A_{\rm ex}$ is well reproduced by Cm_Q^{-3} rather than Cm_Q^{-2}, which implies that the quark-mass dependence of repulsion seems stronger than that of the strength of the CM interaction itself. However, considering that the Pauli blocking among quarks is essential for the repulsion and that the interaction range of the repulsive force has a very weak quark-mass dependence, the present results are consistent with the quark-model interpretation that short-range repulsion between hadrons arises from CM interaction among quarks.

5. Summary and Outlooks

We have evaluated inter-hadron interactions in two-color lattice QCD. We have considered two-diquark scatterings, whose flavors are (i,j) and (k,l), and extracted its interhadron potential $V_{ij,kl}(R)$. We have found

- *The existence of universal attraction between two hadrons.*
- *The measure of Pauli-blocking effect,* $V_{\rm ex}(R) \equiv V_{12,12}(R) - V_{12,34}(R)$, *always shows repulsive contribution at any* κ*'s.*
- *The quark-mass dependence of the strength of* $V_{\rm ex}$ *is consistent with or stronger than* m_Q^{-2}.
- *The quark-mass dependences of the interaction ranges of direct and exchange parts of potentials (*$V_{\rm dir}$,$V_{\rm ex}$*) are much weaker than expected from meson exchanges.*

Since quark-mass dependence in interaction ranges does not appear, the origin of the interactions we observed so far would be all gluonic interactions and/or flavor-exchange processes, *e.g.* color-magnetic or color-Coulomb interactions, and so on, rather than mesonic contributions. As was found in our analyses, attractive forces are readily masked by the strong repulsive force in $V_{12,12}$ in the light quark-mass region. If the universal attraction in hadronic interaction appears also in SU(3) QCD, they might be observed in a channel with no or less quark-exchange contribution between hadrons, such as $N\phi$ scattering state. In the lighter quark-mass region, meson-exchange contributions could largely emerge and be predominant, which is left for further studies.

[1] T. T. Takahashi and Y. Kanada-En'yo, Phys. Rev. D **82**, 094506 (2010) [arXiv:0912.0691 [hep-lat]].
[2] R. Jastrow, Phys. Rev. **81**, 165 (1951).
[3] S. Otsuki, M. Yasuno and R. Tamagaki, Prog. Theor. Phys. Suppl. **1965**, 578 (1965).
[4] V. G. Neudachin, Yu. F. Smirnov and R. Tamagaki, Prog. Theor. Phys. **58**, 1072 (1977).
[5] D. A. Liberman, Phys. Rev. D **16**, 1542 (1977).
[6] M. Harvey, Nucl. Phys. A **352**, 326 (1981).
[7] M. Oka and K. Yazaki, Phys. Lett. B **90**, 41 (1980); Prog. Theor. Phys. **66**, 556 (1981); Prog. Theor. Phys. **66**,572 (1981).
[8] A. Faessler, F. Fernandez, G. Lubeck and K. Shimizu, Phys. Lett. B **112**, 201 (1982).
[9] M. Oka and K. Yazaki, in *Quarks and nuclei*, ed. W. Weise (World Scientific, Singapore, 1984), p. 489 and references theirein .
[10] M. Oka, K. Shimizu and K. Yazaki, Prog. Theor. Phys. Suppl. **137**, 1 (2000).
[11] N. Ishii, S. Aoki and T. Hatsuda, Phys. Rev. Lett. **99**, 022001 (2007) [arXiv:nucl-th/0611096].
[12] S. Aoki *et al.* [CP-PACS Collaboration], Phys. Rev. D **71**, 094504 (2005) [arXiv:hep-lat/0503025].
[13] K. Sasaki and N. Ishizuka, Phys. Rev. D **78**, 014511 (2008) [arXiv:0804.2941 [hep-lat]].
[14] H. Nemura, N. Ishii, S. Aoki and T. Hatsuda, Phys. Lett. B **673**, 136 (2009) [arXiv:0806.1094 [nucl-th]].
[15] Y. Ikeda and H. Iida, arXiv:1102.2097 [hep-lat].
[16] Y. Koike, O. Morimatsu and K. Yazaki, Nucl. Phys. A **449**, 635 (1986).

Bound-state properties from field-theory correlators

Dmitri Melikhov

HEPHY, Austrian Academy of Sciences, Nikolsdorfergasse 18, A-1050 Vienna, Austria
Faculty of Physics, University of Vienna, Boltzmanngasse 5, A-1090 Vienna, Austria
SINP, Moscow State University, 119991, Moscow, Russia

E-mail: dmitri_melikhov@gmx.de

Abstract. We discuss the details of calculating hadron properties from the OPE for correlators of quark currents in QCD, which constitutes the basis of the method of QCD sum rules. The main emphasis is laid on gaining control over the systematic uncertainties of the hadron parameters obtained within this method. We start with examples from quantum mechanics, where bound-state properties may be calculated independently in two ways: exactly, by solving the Schrödinger equation, and approximately, by the method of sum rules. Knowing the exact solution allows us to control each step of the sum-rule extraction procedure. On the basis of this analysis, we formulate several improvements of the method of sum rules. We then apply these modifications to the analysis of the decay constants of heavy mesons.

1. Introduction

The method of dispersive sum rules [1] is one of the widely used methods for obtaining properties of ground-state hadrons in QCD. The method involves two steps: (i) one calculates the relevant correlator in QCD at relatively small values of the Eucledian time; (ii) one applies numerical procedures suggested by quark-hadron duality in order to isolate the ground-state contribution from this correlator. These numerical procedures cannot determine a single value of the ground-state parameter but should provide the band of values containing the true hadron parameter. This band is a systematic, or intrinsic, uncertainty of the method of sum rules.

An unbiased judgement of the reliability of the extraction procedures adopted in the method of sum rules may be acquired by applying these procedures to problems where the ground-state parameters may be found independently and exactly as soon as the parameters of theory are fixed. Presently, only quantum-mechanical potential models provide such a possibility: (i) the bound-state parameters (masses, wave functions, form factors) are known precisely from the Schrödinger equation; (ii) direct analogues of the QCD correlators may be calculated exactly.

Making use of these models, we studied the extraction of ground-state parameters from different types of correlators: namely, the ground-state decay constant from two-point vacuum-to-vacuum correlator [2], the form factor from three-point vacuum-to-vacuum correlator [3], and the form factor from vacuum-to-hadron correlator [4]. We have demonstrated that the standard procedures adopted in the method of sum rules not always work properly: the true value of the bound-state parameter was shown to lie outside the band obtained according to the standard criteria. These results gave us a solid ground to claim that also in QCD the actual accuracy of the method may be worse than the accuracy expected on the basis of applying the standard criteria.

We realized that the main origin of these problems of the method originate from an over-simplified model for hadron continuum which is described as a perturbative contribution above a constant Borel-parameter independent effective continuum threshold. We introduced the notion of the *exact* effective continuum threshold, which corresponds to the true bound-state parameters: in potential models the true hadron parameters—decay constant and form factor—are known and the exact effective continuum thresholds for different correlators may be calculated. We have demonstrated that the exact effective continuum threshold (i) is not a universal quantity and depends on the correlator considered (i.e., it is in general different for two-point and three-point vacuum-to-vacuum correlators), (ii) depends on the Borel parameter and, for the form-factor case, also on the momentum transfer [3, 4, 5].

In recent publications [6] we proposed a new algorithm for extracting the parameters of the ground state. The idea formulated in these papers is to relax the standard assumption of a Borel-parameter independent effective continuum threshold and to allow for a Borel-parameter dependent quantity. This lecture explains the details of this procedure and its application to decay constants of heavy mesons.

2. OPE and sum rule in quantum-mechanical potential model

Let us start with quantum-mechanical potential model described by Hamiltonian

$$H = H_0 + V(r); \qquad H_0 = \frac{k^2}{2m}, \qquad V(r) = V_{\rm conf}(r) - \frac{\alpha}{r}. \tag{1}$$

Polarization operator $\Pi(E)$ and its Borel transform $\Pi(T)$ [$E \to T$, $1/(H-E) \to \exp(-HT)$, T the Borel parameter] are defined via the full Green function $G(E) = 1/(H-E)$ [7]:

$$\Pi(E) = \langle \vec{r}_f = 0 | \frac{1}{H-E} | \vec{r}_i = 0 \rangle, \qquad \Pi(T) = \langle \vec{r}_f = 0 | \exp(-HT) | \vec{r}_i = 0 \rangle. \tag{2}$$

The expansion of $G(E)$ and $\Pi(E)$ in powers of the interaction is obtained with the help of the Lippmann-Schwinger equation

$$G(E) = G_0(E) - G_0(E) V G_0(E) + G_0(E) V G_0(E) V G_0(E) + \dots, \tag{3}$$

where $G_0(E) = 1/(H_0 - E)$. Since the interaction contains now two parts, $V_{\rm conf}(r)$ and $\frac{\alpha}{r}$, the expansion (3) is a double expansion in powers of $V_{\rm conf}$ and α. E.g., for the case $V_{\rm conf}(r) = \frac{m\omega^2 r^2}{2}$ one easily obtains the corresponding double expansion in powers of α and ωT:

$$\begin{aligned}
\Pi_{\rm OPE}(T) &= \Pi_{\rm pert}(T) + \Pi_{\rm power}(T), \quad \Pi_{\rm pert}(T) = \left(\frac{m}{2\pi T}\right)^{3/2} \left[1 + \sqrt{2\pi m T}\alpha + \frac{1}{3} m\pi^2 T \alpha^2\right], \\
\Pi_{\rm power}(T) &= \left(\frac{m}{2\pi T}\right)^{3/2} \left[-\frac{1}{4}\omega^2 T^2 \left(1 + \frac{11}{12}\sqrt{2\pi m T}\alpha\right) + \frac{19}{480}\omega^4 T^4\right].
\end{aligned} \tag{4}$$

One can see here a "perturbative contribution" (i.e. not containing the confining potential), and "power corrections" given in terms of the confining potential (including also mixed terms containing contributions from both Coulomb and confining potentials). A perturbative contribution may be written in the form of spectral representation [8] yielding

$$\Pi_{\rm OPE}(T) = \int_0^\infty dz e^{-zT} \rho_{\rm pert}(z) + \Pi_{\rm power}(T), \; \rho_{\rm pert} = \left(\frac{m}{2\pi}\right)^{3/2} \left[2\sqrt{\frac{z}{\pi}} + \sqrt{2\pi m}\alpha + \frac{\pi^{3/2} m \alpha^2}{3\sqrt{z}}\right] \tag{5}$$

The "physical" representation for $\Pi(T)$—in the basis of hadron eigenstates—reads:

$$\Pi_{\rm phys}(T) = \langle \vec{r}_f = 0|\exp(-HT)|\vec{r}_i = 0\rangle = \sum_{n=0}^{\infty} R_n \exp(-E_n T), \quad R_n = |\Psi_n(\vec{r}=0)|^2. \tag{6}$$

Sum rule is the expression of the fact that the correlator may be calculated in two ways—using the basis of quark states (OPE) or confined bound states—leading to the same result:

$$\Pi_{\rm OPE}(T) = \Pi_{\rm phys}(T). \tag{7}$$

In order to isolate the ground-state contribution one needs the information about the excited states. A standard Ansatz for the hadron spectral density has the form [1]

$$\rho_{\rm phys}(z) = R_g \delta(z - E_g) + \theta(z - z_{\rm eff})\rho_{\rm pert}(z). \tag{8}$$

It assumes that the contribution of the excited states may be described by contributions of diagrams of perturbation theory above some effective continuum threshold $z_{\rm eff}$. This effective continuum threhsold (different from the physical continuum threhsold which is determined by hadron masses) is an additional parameter of the method of sum rules. Using Eq. (8) yields

$$R_g e^{-E_g T} = \int_0^{z_{\rm eff}} dz e^{-zT} \rho_{\rm pert}(z) + \Pi_{\rm power}(T) \equiv \Pi_{\rm dual}(T, z_{\rm eff}). \tag{9}$$

As soon as one knows $z_{\rm eff}$, one immediately obtains estimates for R_g and E_g, $R_{\rm dual}(T, z_{\rm eff})$ and $E_{\rm dual}(T, z_{\rm eff}) = -d_T \log \Pi(T, z_{\rm eff})$. These however depend on unphysical parameters T and $z_{\rm eff}$.

Eq. (8) is motivated by quark-hadron duality which claims that far above the threshold the hadron spectral density is well described by diagrams of perturbation theory. However, near the physical threshold—and this very region turns out to be essentail for the calculation of ground-state properties—the duality relation is violated.

The advantage of quantum mechanics is that the exact E_g and R_g may be obtained by solving Schrödinger equation [9]. We can then calculate $z_{\rm eff}$ from

$$R_g e^{-E_g T} = \int_0^{z_{\rm eff}} dz e^{-zT} \rho_{\rm pert}(z) + \Pi_{\rm power}(T). \tag{10}$$

The obtained "exact threshold" $z_{\rm eff}(T)$ is a slightly rising function of T [2].

3. OPE and sum rule in QCD

In QCD, the sum rule for the decay constants of heavy mesons has a similar form [10]

$$f_Q^2 M_Q^4 e^{-M_Q^2 \tau} = \int_{(m_Q+m_u)^2}^{s_{\rm eff}} e^{-s\tau} \rho_{\rm pert}(s, \alpha, m_Q, \mu)\, ds + \Pi_{\rm power}(\tau, m_Q, \mu) \equiv \Pi_{\rm dual}(\tau, \mu, s_{\rm eff}) \tag{11}$$

In order to extract the decay constant one should fix the effective continuum threshold $s_{\rm eff}$ which should be a function of τ; otherwise the τ-dependences of the l.h.s. and the r.h.s. of (11) do not match each other. The exact $s_{\rm eff}$ corresponding to the exact hadron decay constant and mass on the l.h.s. is of course not known. The extraction of hadron parameters from the sum rule consists therefore in attempting (i) to find a good approximation to the exact threshold and (ii) to control the accuracy of this approximation. For further use, we define the dual decay constant $f_{\rm dual}$ and the dual invariant mass $M_{\rm dual}$ by relations

$$f_{\rm dual}^2(\tau) = M_Q^{-4} e^{M_Q^2 \tau} \Pi_{\rm dual}(\tau, s_{\rm eff}(\tau)), \quad M_{\rm dual}^2(\tau) = -\frac{d}{d\tau} \log \Pi_{\rm dual}(\tau, s_{\rm eff}(\tau)). \tag{12}$$

4. Extraction of the decay constant: QCD vs. potential model [11]

If the mass of the ground state is known, the deviation of the dual mass from the actual mass of the ground state gives an indication of the contamination of the dual correlator by excited states. Assuming a specific functional form of the effective threshold and requiring the least deviation of the dual mass (12) from the known ground-state mass in the τ-window leads to a variational solution for the effective threshold. As soon as the latter has been fixed, one calculates the decay constant from (12). **Our algorithm for extracting ground-state parameters reads:**
(i) Consider a set of Polynomial τ-dependent Ansaetze for $s_{\rm eff}$: $s^{(n)}_{\rm eff}(\tau) = s^{(n)}_0 + s^{(n)}_1\tau + s^{(n)}_2\tau^2 + \dots$.
(ii) Calculate the dual mass for these τ-dependent thresholds and minimize the squared difference between the "dual" mass $M^2_{\rm dual}$ and the known value M^2_B in the τ-window. This gives the parameters of the effective continuum thresholds $s^{(n)}_i$.
(iii) Making use of the obtained thresholds, calculate the decay constant.
(iv) Take the band of values provided by the results corresponding to **linear**, **quadratic**, and **cubic** effective thresholds as the characteristic of the intrinsic uncertainty of the extraction procedure.

Figure 1 shows the application of our algorithm in quantum mechanics and in QCD. In both cases a very similar pattern emerges; the reproduction of the dual mass considerably improves for the τ-dependent quantities indicating that the dual correlator with a τ-dependent threshold isolates the contribution of the ground state much better than the dual correlator with a standard τ-independent threshold. As a consequence, the accuracy of the extracted hadron observable improves considerably.

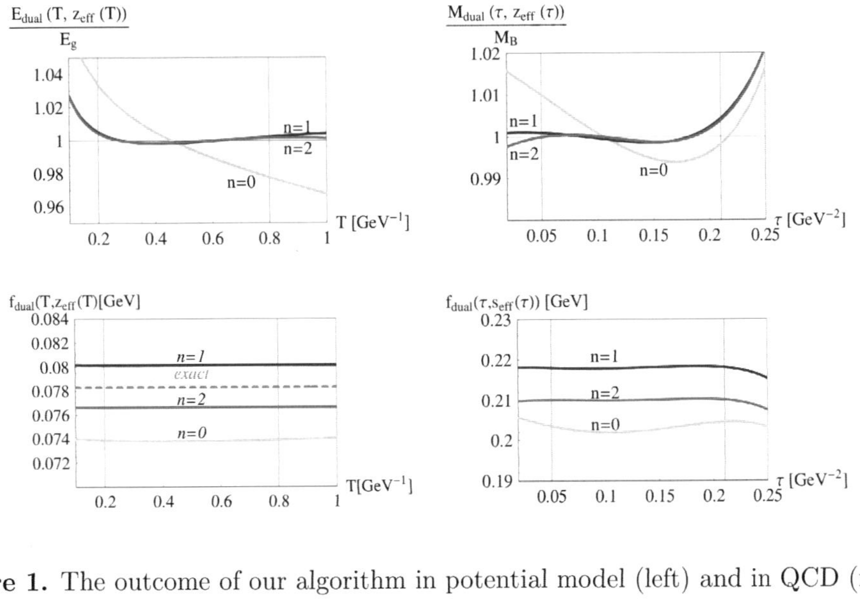

Figure 1. The outcome of our algorithm in potential model (left) and in QCD (right).

5. Decay constants of D and D_s mesons [12]

The application of our extraction procedures leads to the following results:

$$f_D = 206.2 \pm 7.3_{\rm (OPE)} \pm 5.1_{\rm (syst)}\,{\rm MeV}, \qquad f_{D_s} = 245.3 \pm 15.7_{\rm (OPE)} \pm 4.5_{\rm (syst)}\,{\rm MeV}. \qquad (13)$$

The τ-dependent threshold is a crucial ingredient for a successful extraction of the decay constant from the sum rule (Fig. 2). Obviously, the standard τ-independent approximation leads to much lower value for f_D which lies rather far from the data and the lattice results.

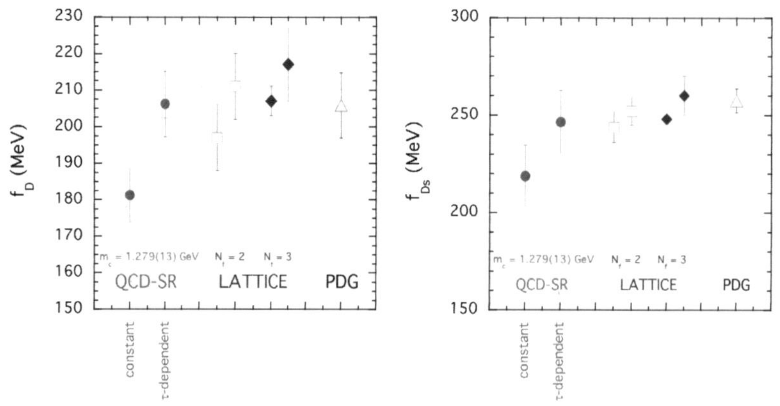

Figure 2. Our results for f_D and f_{Ds} [12].

6. Conclusions

The effective continuum threshold $s_{\rm eff}$ is an important ingredient of the method of dispersive sum rules which determines to a large extent the numerical values of the extracted hadron parameter. Finding a criterion for fixing $s_{\rm eff}$ poses a problem in the method of sum rules.

• $s_{\rm eff}$ depends on the *external kinematical variables* (e.g., momentum transfer in sum rules for 3-point correlators and light-cone sum rules) and *''unphysical" parameters* (renormalization scale μ, Borel parameter τ). Borel-parameter τ-dependence of $s_{\rm eff}$ emerges naturally when trying to make quark-hadron duality more accurate.

• We proposed a new algorithm for fixing τ-dependent $s_{\rm eff}$, for those problems where the bound-state mass is known. We have tested that our algorithm leads to more accurate values of ground-state parameters than the "standard" algorithms used in the context of dispersive sum rules before. Moreover, our algorithm allows one to probe *"intrinsic"* uncertainties related to the limited accuracy of the extraction procedure in the method of QCD sum rules.

Acknowledgments. I have pleasure to thank Wolfgang Lucha and Silvano Simula for the most pleasant and fruitful collaboration on the subject presented in this lecture and to the Austrian Science Fund FWF for support under Project No. P20573. Special thanks are due to Organizers for creating a stimulating and pleasant atmosphere of this Conference.

References

[1] M. Shifman, A. Vainshtein, and V. Zakharov, Nucl. Phys. **B147**, 385 (1979).
[2] W. Lucha, D. Melikhov, and S. Simula, Phys. Rev. **D76**, 036002 (2007); Phys. Lett. **B657**, 148 (2007); Phys. Atom. Nucl. **71**, 1461 (2008).
[3] W. Lucha and D. Melikhov, Phys. Rev. **D73**, 054009 (2006); Phys. Atom. Nucl. **70**, 891 (2007). W. Lucha, D. Melikhov, and S. Simula, Phys. Lett. **B671**, 445 (2009).
[4] W. Lucha, D. Melikhov, and S. Simula, Phys. Rev. **D75**, 096002 (2007); Phys. Atom. Nucl. **71**, 545 (2008). D. Melikhov, Phys. Lett. **B671**, 450 (2009).
[5] I. Balakireva, W. Lucha, and D. Melikhov, arXiv:1103.3781, arXiv:1110.2080, arXiv:1110.6904.
[6] W. Lucha, D. Melikhov, and S. Simula, Phys. Rev. **D79**, 096011 (2009); J. Phys. **G37**, 035003 (2010). W. Lucha, D. Melikhov, H. Sazdjian, and S. Simula, Phys. Rev. **D80**, 114028 (2009).
[7] A. Vainshtein, V. Zakharov, V. Novikov, M. Shifman, Sov. J. Nucl. Phys. **32**, 840 (1980).
[8] M. Voloshin, Int. J. Mod. Phys. A **10**, 2865 (1995).
[9] W. Lucha and F. Schöberl, Int. J. Mod. Phys. C **10**, 607 (1999).
[10] T. M. Aliev and V. L. Eletsky, Yad. Fiz. **38**, 1537 (1983);
M. Jamin and B. O. Lange, Phys. Rev. **D65**, 056005 (2002).
[11] W. Lucha, D. Melikhov, and S. Simula, Phys. Lett. **B687**, 48 (2010); Phys. Atom. Nucl. **73**, 1770 (2010).
[12] W. Lucha, D. Melikhov, and S. Simula, Phys. Lett. **B701**, 82 (2011); J. Phys. **G38**, 105002 (2011).

Improved sum rules for light mesons and thermal hadronic threshold

E Veli Veliev*, **A Türkan, N Akşit**

Department of Physics, Kocaeli University, 41380 Izmit, Turkey

E-mail: *elsen@kocaeli.edu.tr

Abstract. The thermal properties of light vector and scalar mesons are investigated in the framework of QCD sum rules. The phenomenological side of the correlation function can be calculated using either the quark-hadron duality approach or in terms of two-pion continuum contributions. In the quark-hadron duality approach, a free parameter (hadronic threshold) arises in the phenomenological part and it is necessary to know the temperature dependence of the hadronic threshold. A comparison of above mentioned approaches gives us additional information about the temperature dependence of the hadronic threshold. Taking into account the thermal spectral density and additional operators appearing at finite temperature the thermal QCD sum rules are improved. The decay constants of ρ and σ mesons are calculated and our investigations show that the above mentioned methods give us the same results.

1. Introduction

The nature of light mesons is a long-standing problem in hadron physics for several decades. In spite of the striking success of QCD theory for strong interaction, the underlying structure of the light mesons is still not fully understood. Especially the light scalar mesons have the same quantum numbers $J^{PC} = 0^{++}$ as the vacuum and so their structure is major issue in order to understand non-perturbative properties of the QCD vacuum.

In the literature there are various approaches and models which predict the thermal behavior of hadrons. The QCD sum rules method is one of the powerful methods for investigating the properties of hadrons [1]. QCD sum rules, which is based on the operator product expansion (OPE), QCD Lagrangian and quark-hadron duality, was later extended to the finite temperature case [2]. However, extension of sum rules to finite temperature case have some complications [3-5].

The leptonic decay constants of ρ and σ mesons are widely investigated in the studies [6,7]. In these studies, two-pion continuum contributions are used to obtain decay constants. But the quark-hadron duality approach requires knowledge on the temperature dependence of hadronic threshold, which generally is expressed by quark condensate $\langle\bar{\psi}\psi\rangle$ in the following form [8]:

$$s_0(T) = s_0\frac{\langle\bar{\psi}\psi\rangle}{\langle 0|\bar{\psi}\psi|0\rangle}\Big(1-\frac{4m^2}{s_0}\Big)+4m^2. \tag{1}$$

where s_0 is zero temperature hadronic threshold and m is quark mass.

In this study we reanalyze the decay constants of ρ and σ mesons and compare our results with the results existing in the literature [6,7]. We take into account the thermal spectral density and perturbative two-loop order corrections to the correlation function and improve thermal QCD sum rules obtained in [6,7]. The leptonic decay constants are calculated using the values of energy density, quark and gluon condanstes obtained via chiral perturbation theory [9] and lattice QCD theory [10,11]. Our calculations show that obtained results using both two pion continuum approach and quark-hadron duality approach are compatible with each other.

2. Improved QCD sum rules and thermal hadronic threshold

In this section we reanalyze the thermal sum rules for ρ and σ mesons. For this aim we start with two-point thermal correlation function,

$$\Pi\Big(q,T\Big) = i\int d^4x\ e^{iq\cdot x} Tr\Big(\rho\ \mathcal{T}\Big(J(x)J^\dagger(0)\Big)\Big), \tag{2}$$

where $\rho = e^{-\beta H}/Tre^{-\beta H}$ is the thermal density matrix of QCD at temperature $T = 1/\beta$, $\mathcal{T}$ indicates the time ordered product and $J(x) =: \overline{Q}(x)\Gamma Q(x) :$ is interpolating current. In interpolating current $Q(x)$ is quark field and $\Gamma = I$ or γ_μ for scalar and vector particles, respectively. The thermal spectral densities in lowest order of perturbation theory are calculated in the papers [2,12] and they are given by

$$\rho(s,T) \ = \ \frac{s}{8\pi^2}v(s)f(s)\Big(1-2\,n\Big(\frac{\sqrt{s}}{2}\Big)\Big), \tag{3}$$

where $v(s) = \sqrt{1-4m^2/s}$ and $f(s) = 3v^2(s)$ or $f(s) = 3-v^2(s)$ for σ and ρ mesons, respectively.

In our calculations, we also take into account the perturbative two-loop order α_s correction to the spectral densities. These corrections at zero temperature can be written as [13]:

$$\rho_{\alpha_s}(s,T) = \frac{11s}{8\pi^3}\alpha_s \qquad and \qquad \rho_{\alpha_s}(s,T) = \frac{s}{4\pi^3}\alpha_s, \tag{4}$$

for σ and ρ mesons, respectively. In Eqs. (4) we replace the strong coupling α_s with its temperature dependent lattice improved expression $\alpha(T) = 2.095(82)g^2(T)/4\pi$ (for more details see [12]).

To calculate the phenomenological side of correlation function, we insert a complete set of physical intermediate state to Eq. (2). The decay constants f_σ and f_ρ are defined by the matrix element of the scalar current $J(0)$ and vector current $J_\mu(0)$ between the vacuum and hadronic states:

$$\langle 0|J(0)|\sigma\rangle = f_\sigma m_\sigma, \qquad and \qquad \langle 0|J_\mu(0)|\rho\rangle = f_\rho m_\rho \varepsilon^\lambda_\mu, \tag{5}$$

where ε^λ_μ is polarization states of vector mesons. Finally matching the spectral representation and results of OPE and using quark-hadron duality, our improved sum rules both σ and ρ-mesons take the form:

$$f^2(T)\ m^2(T)\ \exp\Big(-\frac{m^2(T)}{M^2}\Big) = \int_{4m^2}^{s_0(T)} ds\ [\rho(s)+\rho_{\alpha_s}(s)]\exp\Big(-\frac{s}{M^2}\Big) + \hat{B}\Pi^{np}(q,T), \tag{6}$$

where M^2 is Borel mass parameter. In Eq. (6) the non-perturbative parts for σ and ρ-mesons are given by [6,7]

$$\hat{B}\Pi^{np}_\sigma\Big(q,T\Big) = 3m\langle\overline{\psi}\psi\rangle + \frac{g^2}{32\pi^2}\langle G^a_{\mu\nu}G^{a\mu\nu}\rangle - \frac{6}{11}\Big[\langle u\Theta u\rangle + \lambda(M^2)\Big(\frac{16}{3}\langle u\Theta^f u\rangle - \langle u\Theta^g u\rangle\Big)\Big], \tag{7}$$

$$\hat{B}\Pi^{np}_\rho(q,T) = -2m\langle\overline{\psi}\psi\rangle - \frac{g^2}{48\pi^2}\langle G^a_{\mu\nu}G^{a\mu\nu}\rangle - \frac{4}{11}\Big(\langle u\Theta u\rangle + \lambda(M^2)\Big(\frac{16}{3}\langle u\Theta^f u\rangle - \langle u\Theta^g u\rangle\Big)\Big), \tag{8}$$

where $\lambda(M^2) = \left(ln\left[\frac{\mu^2}{\Lambda^2}\right]/ln\left[\frac{M^2}{\Lambda^2}\right]\right)^{\frac{44}{87}}$, $\Theta^f_{\mu\nu}$ and $\Theta^g_{\mu\nu}$ are fermionic and gluonic parts of energy momentum tensor $\Theta_{\mu\nu}$, respectively.

Our numerical analysis show that in the intervals $s_0 = (1-1.5)\ GeV^2$ for the σ and $s_0 = (1.5-2)\ GeV^2$ for the ρ channels, the obtained results weakly depend on these parameters. The working region for the Borel mass parameters are $0.3\ GeV^2 \leq M^2 \leq 0.6\ GeV^2$ for σ and $0.5\ GeV^2 \leq M^2 \leq 1.2\ GeV^2$ for ρ mesons, respectively. Taking into account the temperature dependences of continuum threshold, energy density, quark and gluon condensates (for more details see [9-12]) we found that the temperature dependences of decay constants for σ and ρ mesons can be expressed by the following fit function:

$$f = A\exp(-T/\lambda) + B, \tag{9}$$

where the A, λ and B are constant parameters and the obtained values of these parameters are given in Table1.

	$A\ (GeV)$	$\lambda\ (GeV)$	$B(GeV)$
σ	-4.18646×10^{-4}	-0.02988	0.20328
ρ	-4.72423×10^{-5}	-0.01877	0.20966

Table 1. The values of constant parameters in fit function.

We calculated the values of the decay constants of the σ and ρ mesons at $T = 0$ and we found that $f_\sigma = 205.92\ MeV$ and $f_\rho = 210.53\ MeV$. These results are in good consistency with the existing studies in the literature [14-16]. Our investigations show that thermal spectral density and two loop order contributions are significantly important, and two-pion continuum and quark-hadron duality approaches give the same results.

3. Acknowledgement

This work is supported in part by the Scientific and Technological Research Council of Turkey (TUBITAK) under the research project No. 110T284.

References

[1] Shifman M A, Vainstein A I and Zakharov V I 1979 *Nucl. Phys. B* **147** 385 ; Shifman M A, Vainstein A I and Zakharov V I 1979 *Nucl. Phys. B* **147** 448
[2] Bochkarev A I and Shaposhnikov M E 1986 *Nucl. Phys. B* **268** 220
[3] Shuryak E V 1993 *Rev. Mod. Phys.* **65** 1
[4] Hatsuda T, Koike Y and Lee S H 1993 *Nucl. Phys. B* **394** 221
[5] Mallik S 1998 *Phys. Lett. B* **416** 373
[6] Veliev E V and Aliev T M 2008 *J. Phys. G:Nucl. Part. G* Phys. **35** 125002
[7] Mallik S and Mukherjee K 1998 *Phys. Rev. D* **58**, 096011.
[8] Dominguez C A, Loewe M, Rojas J C 2007 *JHEP* **08**, 040
[9] Gerber P and Leutwyler H 1989 *Nucl. Phys. B* **321** 387
[10] Cheng M 2008 *et.al, Phys. Rev. D* **77** 014511
[11] Miller D E 2007 *Phys. Rept.* **443** 55
[12] Veliev E V, Sundu H, Azizi K, Bayar M 2010 *Phys. Rev. D* **82**, 056012
[13] Reinders L J, Rubinstein H and Yazaki S 1985 *Phys. Rep.* **127**, 1
[14] Nakamura K et al. (Particle Data Group) 2010 *J. Phys. G* **37** 075021
[15] Colangelo P and Khodjamirian A 2001 *in At the Frontier of Particle Physics/Handbook of QCD, edited by M. Shifman* vol 3 (Singapore: World Scientific) p 1495
[16] Gokalp A, Yilmaz O 2001 , *Phys. Rev. D* **64**, 034012

Journal of Physics: Conference Series **348** (2012) 012016 doi:10.1088/1742-6596/348/1/012016

Mesons Spectral Functions at Finite Temperature

E Veli Veliev*[1], K Azizi[2], H Sundu[1], G Kaya[1] and N Akşit [1]

[1]Department of Physics, Kocaeli University, 41380 Izmit, Turkey
[2]Physics Division, Faculty of Arts and Sciences, Doğuş University, Acıbadem-Kadıköy, 34722 Istanbul, Turkey

E-mail: *[1]elsen@kocaeli.edu.tr

Abstract. We investigate the thermal spectral densities for (pseudo)scalar and vector currents in the framework of the real time formalism when mass of two quarks are different. Such spectral densities are necessary for the phenomenological investigation of in-medium properties of hadrons. We use the quark propagator at finite temperature and calculate annihilation and scattering parts of spectral densities for above mentioned currents. The investigations show that the thermal contributions are significantly important. The obtained results at $T \rightarrow 0$ limit are in good consistency with the vacuum results.

1. Introduction

One of the most important subjects of the QCD physics is the phenomenological investigation of in-medium modifications of hadrons. Heavy-ion collisions and reactions have been elementarily used to extract experimental information on in-medium properties of hadrons. Also, there are a lot of different theoretical methods which have been widely used to for this purpose in literature. These methods are effective hadronic models, chiral perturbation theory, lattice theory, low-density theorems, quark models and QCD sum rules. QCD sum rules first introduced by Shifman, Vainshtein and Zakharov [1], and its extension version to finite temperature [2] have been one of the most efficient approaches among these methods. QCD sum rule is obtained by equating the QCD and the phenomenological representations of the correlation function and this method establishes a connection between QCD vacuum structure and hadron properties [3]. Determination of correlation function by using dispersion relations is the first step in QCD sum rules and it is necessary to evaluate the spectral density of mentioned current for this aim. Therefore determination of spectral densities is the main tool of QCD sum rules method. Spectral functions in different cases were studied in the literature [4]-[14].

In this paper, we study the thermal spectral densities for different currents at finite temperature. We calculate the annihilation and scattering parts of the spectral densities for (pseudo)scalar and vector currents for $m_1 \neq m_2$ case using the quark propagator in the real time formulation of the thermal field theory. The investigations show that the thermal contributions are significantly important. Also, we show that our obtained results at $T \rightarrow 0$ limit are in good consistency with vacuum results.

2. Thermal spectral densities for various currents

The thermal QCD sum rule approach is based on the evaluation of thermal correlator of the interpolating current $J(x) = \bar{q}_1(x)\Gamma q_2(x)$. Two-point thermal correlation function is given by:

$$\Pi\Big(q,T\Big) = i\int d^4x\, e^{iq\cdot x} Tr\Big(\rho\, \mathcal{T}\Big(J(x)J^\dagger(0)\Big)\Big), \tag{1}$$

where $\mathcal{T}$ denotes the time ordered product and $\rho = e^{-\beta H}/Tre^{-\beta H}$ is the thermal density matrix of QCD at temperature $T = 1/\beta$. In interpolating current $J(x)$, $\Gamma = I$ or $i\gamma_5$ for scalar and pseudoscalar particles, respectively and $\Gamma = \gamma_\mu$ as for vector particles.

Firstly, we consider the thermal spectral density for pseudo(scalar) particles. $\Pi(q,T)$ for these currents can be written in momentum space as:

$$\Pi(q,T) = 4iN_c\int \frac{d^4k}{(2\pi)^4}(k^2 - k\cdot q - m_1m_2)D(k,m_1)D(k-q,m_2), \tag{2}$$

where, $D(k)$ is expressed as $D(k,m) = 1/(k^2 - m^2 + i\varepsilon) + 2\pi i n(|k_0|)\delta(k^2 - m^2)$ and in this propagator $n(x) = [\exp(\beta x) + 1]^{-1}$ is the Fermi distribution function. Carrying out the integral over k_0, we obtain the imaginary part of the $\Pi(q,T)$ as $Im\Pi(q,T) = L(q_0) + L(-q_0)$, where

$$\begin{aligned} L(q_0) &= -N_c\int \frac{d\mathbf{k}}{8\pi^2}\frac{\omega_1^2 - \mathbf{k}^2 + \mathbf{k}\cdot\mathbf{q} - \omega_1 q_0 - m_1m_2}{\omega_1\omega_2}\Big[A(n_1,n_2)\delta(q_0 - \omega_1 - \omega_2) \\ &\quad -B(n_1,n_2)\delta(q_0 - \omega_1 + \omega_2)\Big]. \end{aligned} \tag{3}$$

Here m_1 and m_2 are quark masses, $\omega_1 = \sqrt{\mathbf{k}^2 + m_1^2}$, $\omega_2 = \sqrt{(\mathbf{k}\text{-}\mathbf{q})^2 + m_2^2}$, $n_1 = n(\omega_1)$, $n_2 = n(\omega_2)$, $A(n_1,n_2) = (1-n_1)(1-n_2) + n_1n_2$, $B(n_1,n_2) = (1-n_1)n_2 + (1-n_2)n_1$ and the plus and minus signs in front of m_1, m_2, correspond to the scalar and pseudoscalar particles, respectively. The term, which does not include the Fermi distribution functions, shows the vacuum contribution. Terms including the Fermi distributions depict medium contributions. The delta-functions in the different terms of Eq. (3) control the regions of non-vanishing imaginary parts of $\Pi(q,T)$, which define the position of the branch cuts. Taking into account

$$\delta(q_0 - \omega_1 - \omega_2) = \frac{\omega_2}{|\mathbf{k}||\mathbf{q}|}\delta\Big(\cos\theta + \frac{q^2 - 2q_0\omega_1}{2|\mathbf{k}||\mathbf{q}|}\Big) \tag{4}$$

expression (here θ is angle between of $\mathbf{k}$ and $\mathbf{q}$ momentums) and carrying out some transformations, the annihilation and scattering parts of spectral density is found as:

$$\rho_{a,pert}(s,T) = \rho_0(s)\Big[1 - n\Big(\frac{\sqrt{s}}{2}\Big(1 + \frac{m_1^2 - m_2^2}{s}\Big)\Big) - n\Big(\frac{\sqrt{s}}{2}\Big(1 - \frac{m_1^2 - m_2^2}{s}\Big)\Big)\Big], \tag{5}$$

for $(m_1 + m_2)^2 \leq s \leq \infty$,

$$\rho_{s,pert}(s,T) = \rho_0(s)\Big[n\Big(\frac{\sqrt{s}}{2}\Big(1 + \frac{m_1^2 - m_2^2}{s}\Big)\Big) - n\Big(-\frac{\sqrt{s}}{2}\Big(1 - \frac{m_1^2 - m_2^2}{s}\Big)\Big)\Big], \tag{6}$$

for $0 \leq s \leq (m_1 - m_2)^2$, with $m_1 \geq m_2$. Here, $\rho_0(s)$ is the spectral density in the lowest order of perturbation theory at zero temperature and it is given by

$$\rho_0(s) = \frac{3}{8\pi^2 s}q^2(s)v^n(s), \tag{7}$$

where $q(s) = s - (m_1 - m_2)^2$ and $v(s) = \Big(1 - 4m_1m_2/q(s)\Big)^{1/2}$. Here $n = 3$ and $n = 1$ for scalar and pseudoscalar particles, respectively.

Similarly, we consider thermal spectral density for vector mesons. The correlation function of vector current in thermal field theory is given by Lorentz invariant functions, $\Pi_l\Big(q^2,\omega\Big) = \Pi_2/\overline{q}^2$ and $\Pi_t\Big(q^2,\omega\Big) = -\frac{1}{2}(\Pi_1 + q^2\Pi_2/\overline{q}^2)$. Here $\Pi_1 = g^{\mu\nu}\Pi_{\mu\nu}$, $\Pi_2 = u^\mu\Pi_{\mu\nu}u^\nu$, $\overline{q}^2 = \omega^2 - q^2$, $\omega = u\cdot q$ and u_μ is four-velocity. Carrying out the integral over k_0, we obtain the imaginary part of the $\Pi_2(q,T)$ as $Im\Pi_2 = K(q_0) + K(-q_0)$, where

$$K(q_0) = N_c\int\frac{d\mathbf{k}}{8\pi^2}\frac{\mathbf{k}\cdot\mathbf{q} + q_0\omega_1 - 2\omega_1^2}{\omega_1\omega_2}$$
$$\times\Big[A(n_1,n_2)\delta(q_0-\omega_1-\omega_2) - B(n_1,n_2)\delta(q_0-\omega_1+\omega_2)\Big]. \tag{8}$$

Using Eq. (4) and carrying out the integral over angle θ, the annihilation and scattering parts of thermal spectral densities for vector current at nonzero momentum can be written as:

$$\rho_{l,a}(s,|\mathbf{q}|) = \frac{3}{16\pi^2}\int_{-\nu}^{\nu}dx(1-x^2)\Big[1-2n_+(s,|\mathbf{q}|)\Big], \tag{9}$$

$$\rho_{l,s}(s,|\mathbf{q}|) = \frac{3}{16\pi^2}\int_{\nu}^{\infty}dx(1-x^2)\Big[n_+(s,|\mathbf{q}|) - n_-(s,|\mathbf{q}|)\Big], \tag{10}$$

where $n_+(s,|\mathbf{q}|) = n\Big[\frac{1}{2}(|\mathbf{q}|x) + \sqrt{s}\Big]$ and $n_-(s,|\mathbf{q}|) = n\Big[\frac{1}{2}(|\mathbf{q}|x) - \sqrt{s}\Big]$. For $m_1 = m_2$, zero temperature and $|\mathbf{q}| \to 0$ limit cases, Eqs. (5), (6), (9) and (10) are good consistency with the results existing in the literature [14]-[19]. Also, the investigation of obtained results show that thermal contributions are significantly important and therefore the thermal contributions must be taken into account in analysis of mesons properties in medium and interpretation of heavy ion collision experiments.

3. Acknowledgement

This work is supported in part by the Scientific and Technological Research Council of Turkey (TUBITAK) under the research project No. 110T284 and Research Fund of Kocaeli University under grant no. 2011/029.

References

[1] Shifman M A, Vainstein A I and Zakharov V I 1979 *Nucl. Phys. B* **147** 385 ; Shifman M A, Vainstein A I and Zakharov V I 1979 *Nucl. Phys. B* **147** 448
[2] Bochkarev A I and Shaposhnikov M E 1986 *Nucl. Phys. B* **268** 220
[3] Colangelo P and Khodjamirian A 2001 *in At the Frontier of Particle Physics/Handbook of QCD, edited by M. Shifman* vol 3 (Singapore: World Scientific) p 1495
[4] Rapp R and Wambach J 2000 *Adv. Nucl. Phys.* **25** 1
[5] Leutwyler H and Smilga A 1990 *A Nucl. Phys. B* **342** 302
[6] Wetzorke I, Karsch F, Laermann E, Petreczky P and Stickan S 2002 *Nucl. Phys. B Proc. Suppl.* **106** 510
[7] Aarts G and Martinez Resco J M 2005 *Nucl. Phys. B* **726** 93
[8] van Hees H and Rapp R 2005 *Phys. Lett. B* **606** 59
[9] Sarkar S, Patra B K, Menon V J and Mallik S 2002 *Indian J.Phys.* **76A** 385
[10] Muehlich P, Shklyar V, Leupold S, Mosel U, Post M, 2006 *Nucl.Phys. A* **780** 187
[11] Aarts G, Allton C, Foley J, Hands S, Kim S, 2007 *Nucl.Phys. A* **785** 202
[12] Mallik S, Sarkar S 2009 *Eur.Phys. J.C* **61**489
[13] Ghosh S, Sarkar S and Mallik S 2010 *Phys. Rev. C* **82** 045202
[14] Veliev E V and Kaya G 2010 *Acta Phys. Polon. B* **41** 1905
[15] Veliev E V, Azizi K, Sundu H and Akşit N 2011 *arXiv:hep-ph/1010.3110 J. Phys. G* to be published.
[16] Mallik S and Mukherjee K 1998 *Phys. Rev. D* **58**, 096011
[17] Dominguez C A, Loewe M and Rojas J C 2007 *JHEP* **08** 040
[18] Dominguez C A, Loewe M, Rojas J C and Zhang Y 2011 *Phys. Rev. D* **83** 034033
[19] Veliev E V, Sundu H, Azizi K and Bayar M 2010 *Phys. Rev. D* **82** 056012

Different Contributions of the Decay $\rho \to \pi^0\pi^0\gamma$

Esra Aciksoz and Ayse Kucukarslan

Canakkale Onsekiz Mart University, Department of Physics, Canakkale, Turkey, 17100

E-mail: akucukarslan@comu.edu.tr

Abstract. The radiative decay $\rho \to \pi^0\pi^0\gamma$ is re-examined within the framework of a phenomenological approach in which the f_2 meson and σ are combined with the processes involving the vector meson intermediate states and the pion-loops. In order to obtain the scalar σ meson contribution we consider pion-loop model and no structure model, and then compare the results obtained in two models. Furthermore, using the experimental branching ratio for the radiative decay $\rho \to \pi^0\pi^0\gamma$, we estimate the coupling constant $g_{\rho\sigma\gamma}$ for a set of values of σ meson parameters M_σ and Γ_σ.

1. Introduction

The radiative decays of low-mass vector mesons into a pair of neutral pseudoscalars are useful tool for investigating the nature and extracting the properties of the scalar meson resonances. One of these decays which is the decay $\rho \to \pi^0\pi^0\gamma$ should be considered as an ideal process for analyzing the controversial and not well established scalar meson σ. For the radiative decay $\rho \to \pi^0\pi^0\gamma$ there are a number of experimental and theoretical studies. The recent measured branching ratio of the decay $\rho \to \pi^0\pi^0\gamma$ is BR($\rho \to \pi^0\pi^0\gamma$) $= 4.5 \pm 0.8\text{x}10^{-5}$ [1]. On the theoretical side, this decay is studied using the vector meson dominance (VMD) model [2], the chiral perturbation theory [3], and the unitarized chiral perturbation theory (UChPT) [4,5].

2. Formalism

In this work, we follow a phenomenological approach which is also used to the similar calculations in [6,7] and attempt to calculate the decay rate for the decay $\rho \to \pi^0\pi^0\gamma$. We assume that the radiative decay $\rho \to \pi^0\pi^0\gamma$ proceeds through the reactions $\rho \to \omega\pi \to \pi^0\pi^0\gamma$, $\rho \to \sigma\gamma \to \pi^0\pi^0\gamma$, $\rho \to f_2\gamma \to \pi^0\pi^0\gamma$ and $\rho \to \pi^+\pi^-\gamma \to \pi^0\pi^0\gamma$. For the contribution of the scalar σ meson, we use both the pion-loop model, and the no-structure model.

3. Results

We also estimate the coupling constant $g_{\rho\sigma\gamma}$ depending on a set of values of the σ meson parameters [8,9], as shown in table 1. Our estimations are in general quite different than the values existing in the literature, such as $g_{\rho\sigma\gamma} = 2.71$ [10], $g_{\rho\sigma\gamma} = 3.2\pm0.6$ [11] and $g_{\rho\sigma\gamma} = 2.2\pm0.4$ [12]. We also obtain for the branching ratios for different contributing reactions to the radiative decay $\rho \to \pi^0\pi^0\gamma$ as BR($\rho \to \omega\pi \to \pi^0\pi^0\gamma$)=1.15x$10^{-5}$, BR($\rho \to \pi^+\pi^-\gamma \to \pi^0\pi^0\gamma$)=1.07x$10^{-5}$, BR($\rho \to f_2\gamma \to \pi^0\pi^0\gamma$)=1.38x$10^{-6}$ for $G_{f\pi\pi}$=3.14 and BR($\rho \to f_2\gamma \to \pi^0\pi^0\gamma$)=1.712x$10^{-5}$ for $G_{f\pi\pi}$=5.89. For the contribution of the σ, the obtained values are shown in table 2 and table 3.

Table 1. The values of the coupling constants, $g_{\sigma\pi\pi}$ and $g_{\rho\sigma\gamma}$ depending on the σ parameters.

M_σ (MeV)	Γ_σ(MeV)	$g_{\sigma\pi\pi}$	$g_{\rho\sigma\gamma}$
478	263	4.72	(-0.44) and (3.47)
441	544	7.23	(-0.23)
500	600	6.91	(0.11) and (-2.95)

Table 2. The branching ratios depending on the coupling constant $g_{\rho\sigma\gamma}$ of the decay $\rho \rightarrow \pi^0\pi^0\gamma$ are shown. In each value the predictions without and with the f_2 meson are given with different value of the $G_{f\pi\pi}$.

$g_{\rho\sigma\gamma}$	σ (VMD)	Total (withoutf_2)	Total(withf_2) $G_{f\pi\pi} = 3.14$	Total(withf_2) $G_{f\pi\pi} = 5.89$
-0.44	1.40x10^{-5}	4.35x10^{-5}	4.49x10^{-5}	6.07x10^{-5}
-0.23	0.13x10^{-5}	2.76x10^{-5}	2.89x10^{-5}	4.73x10^{-5}
0.11	0.05x10^{-5}	1.52x10^{-5}	1.66x10^{-5}	3.23x10^{-5}
0.58	0.63x10^{-5}	1.24x10^{-5}	1.38x10^{-5}	2.95x10^{-5}

Table 3. The branching ratios depending on the σ parameters of the decay $\rho \rightarrow \pi^0\pi^0\gamma$ are shown. In each value the predictions without and with the f_2 meson are given with different value of the $G_{f\pi\pi}$.

M_σ (MeV)	Γ_σ (MeV)	σ (pion-loop)	Total (withoutf_2)	Total(withf_2) $G_{f\pi\pi} = 3.14$	Total(withf_2) $G_{f\pi\pi} = 5.89$
478	263	4.17x10^{-5}	4.50x10^{-5}	4.64x10^{-5}	6.21x10^{-5}
478	258	4.13x10^{-5}	4.47x10^{-5}	4.61x10^{-5}	6.18x10^{-5}
441	544	8.88x10^{-5}	8.42x10^{-5}	8.56x10^{-5}	10.13x10^{-5}
800	600	1.80x10^{-5}	1.95x10^{-5}	2.09x10^{-5}	3.66x10^{-5}

References

[1] Nakamura K et al. (Particle Data Group), 2010 *JPG* **37** 075021
[2] Bramon A, Grau A and Pancheri G, 1992 *Phys. Lett.* B **283** 416
[3] Bramon A, Grau A and Pancheri G, 1992 *Phys. Lett.* B **289** 97
[4] Oller J A, 1992 *Phys. Lett.* B **426** 7
[5] Marco E, Hirenzaki S, Oset E and Toki H, 1999 *Phys. Lett.* B **470** 20
[6] Gokalp A , Yilmaz O , 2001 *Phys. Lett.* B **508** 25
[7] Gokalp A , Yilmaz O , 2001 *Phys. Rev.* D **64** 053017
[8] Achasov M N et al. 2002 *Phys. Lett.* B **537** 201
[9] Caprini I, Colangelo G and Leutwyler H, 2006 *Phys. Rev. Lett.* **96** 132001
[10] Friman B, Soyeur M, 1996 *Nucl. Phys.* A **600** 477
[11] Gokalp A , Yilmaz O , 2001 *Phys. Rev.* D **64** 053017

Scalar Mesons in the Decay $\phi \rightarrow K^0 \bar{K^0} \gamma$

Yasemin Unal and Ayse Kucukarslan

Canakkale Onsekiz Mart University, Department of Physics, Canakkale, Turkey, 17100

E-mail: akucukarslan@comu.edu.tr

Abstract. We estimate the branching ratio of the $\phi \rightarrow K^0 \bar{K^0} \gamma$ decay within the framework of a phenomenological approach in which the contributions of the $f_0(980)$, $a_0(980)$ scalar resonances, chiral loops and $K^*(892)$ vector meson are considered. To evaluate the contribution of the scalar resonances, we consider the kaon-loop model and the no-structure model.

1. Introduction

The $\phi \rightarrow K^0 \bar{K^0} \gamma$ decay is interesting process to study because it is the only radiative decay where the f_0 and a_0 scalar mesons contribute simultaneously. This will allow to provide us valuable information on the properties of both scalar mesons. On the other side, this process could pose a background problem for testing CP-violation. The existing experimental result for the branching ratio of this decay is BR($\phi \rightarrow K^0 \bar{K^0} \gamma$)$< 1.9 \times 10^{-8}$ [1]. Some theoretical predictions found in literature are BR($\phi \rightarrow K^0 \bar{K^0} \gamma$)$= 1.3 \times 10^{-8}$ [2], BR($\phi \rightarrow K^0 \bar{K^0} \gamma$)$= 7.6 \times 10^{-9}$ [3], BR($\phi \rightarrow K^0 \bar{K^0} \gamma$)$= 5.0 \times 10^{-8}$ [4] BR($\phi \rightarrow K^0 \bar{K^0} \gamma$)$= 7.5 \times 10^{-8}$ [5], BR($\phi \rightarrow K^0 \bar{K^0} \gamma$)$= 4.47 \times 10^{-8}$ [6] for the $\phi \rightarrow K^0 \bar{K^0} \gamma$ decay.

2. Formalism

In this work, we attempt to study the radiative $\phi \rightarrow K^0 \bar{K^0} \gamma$ decay considering the kaon-loop model and no structure model for the coupling of the scalar mesons in the intermediate state to the initial vector meson. We calculate the branching ratios of the different contributions, chiral loops and K^* meson of this decay in addition to the mechanisms of the scalar mesons in the phenomenological approach. In each model of the scalar mesons, we calculate the branching ratio using the different values of the scalar meson parameters and then compare the results.

3. Results and Discussion

The values we obtain for the branching ratio in two models are shown in table 1 and table 2. We present the role of the f_0 and a_0 scalar mesons in the reaction mechanism of the radiative $\phi \rightarrow K^0 \bar{K^0} \gamma$ decay in tables. Hence, the f_0 meson contribution one order stronger than the a_0 meson contribution in the structure model. Considering the total branching ratios, our predictions in two models are not different from each other. Our values obtained here are 1 order of magnitude larger than the experimental result [1] and other theoretical predictions [2,3,4,5,6].

In this work we also determine the total contributions of the scalar mesons, f_0 and a_0, and their interference as BR($\phi \rightarrow K^0 \bar{K^0} \gamma$)=$4.20 \times 10^{-8}$ and BR(int.)=-2.96×10^{-7}, BR($\phi \rightarrow K^0 \bar{K^0} \gamma$)=$4.73 \times 10^{-8}$ and BR(int.)=-5.87×10^{-7} and BR($\phi \rightarrow K^0 \bar{K^0} \gamma$)=$8.73 \times 10^{-8}$

and BR(int.)=-6.20×10^{-7} in the kaon loop model, and BR($\phi \to K^0\bar{K^0}\gamma$)=$2.42 \times 10^{-9}$ and BR(int.)=1.11×10^{-6}, BR($\phi \to K^0\bar{K^0}\gamma$)=$5.80 \times 10^{-8}$ and BR(int.)=1.99×10^{-6} and BR($\phi \to K^0\bar{K^0}\gamma$)=$5.92 \times 10^{-9}$ and BR(int.)=1.78×10^{-6} in the no structure model. For the contributions of the chiral loops and K^{*0} vector meson, we obtain the branching ratios as BR($\phi \to K^0\bar{K^0}\gamma$)=$4.92 \times 10^{-8}$ and BR($\phi \to K^0\bar{K^0}\gamma$)=$2.91 \times 10^{-11}$, respectively. The value obtained from chiral loops is in agreement with estimate in [5]. The exchange of intermediate K^{*0} vector meson is one order of magnitude larger than the value in [5] but it is still small and therefore negligible.

Table 1. The branching ratios of the $\phi \to K^0\bar{K^0}\gamma$ decay depending on different values of the coupling constants $g_{f_0K^+K^-}$ and $g_{a_0K^+K^-}$. The contributions of the scalar mesons are in the kaon loop model (KLM). Here total contribution includes also chiral loops contribution.

$g_{f_0K^+K^-}(GeV)$	$g_{a_0K^+K^-}(GeV)$	f_0	a_0	Total
[7] 4.021 ± 0.01	[7] 2.747 ± 0.42	1.73×10^{-7}	4.85×10^{-8}	2.72×10^{-7}
[7] 5.873	[7] 3.631	2.66×10^{-7}	9.93×10^{-8}	4.17×10^{-7}
[8] 5.919	[9] 2.857	2.71×10^{-7}	5.69×10^{-8}	3.79×10^{-7}

Table 2. The branching ratios of the $\phi \to K^0\bar{K^0}\gamma$ decay depending on different values of the coupling constant $g_{\phi f_0\gamma}$ and $g_{\phi a_0\gamma}$. The contributions of the scalar mesons are in the no structure model (NSM). Here total contribution includes also chiral loops contribution.

$g_{\phi f_0\gamma}(GeV^{-1})$	$g_{\phi a_0\gamma}(GeV^{-1})$	f_0	a_0	Total
[10] 1.48 ± 0.06	[11] 1.61 ± 0.05	4.48×10^{-8}	5.06×10^{-8}	6.68×10^{-7}
[12] 2.78 ± 0.02	[12] 1.83 ± 0.03	7.37×10^{-7}	1.06×10^{-7}	3.10×10^{-6}
[12] 2.61 ± 0.02	[13] 1.30 ± 0.3	8.97×10^{-9}	1.41×10^{-7}	5.43×10^{-7}

References

[1] Ambrosino F, Antonelli A and Antonelliloe M, 2009 Contributed paper to Lepton Photon
[2] Achasov N. N and Ivanchenko V. N, 1989 *Nucl. Phys. B* **315** 465
[3] Bramon A, Grau A and Pancheri G 1992 *Phys. Lett. B* **289** 97
[4] Oller J. A 1998 *Phys. Lett. B* **426** 7
[5] Escribano R 2007 *Eur. Phys. J. A* **31** 454
[6] Gokalp A, Korkmaz C. S and Yilmaz O 2007 *Phys. Rev. D* **75** 013001
[7] Achasov N. N and Gubin V. V, 2001 *Phys. Rev. D* **64** 094016
[8] The KLOE Collaboration, 2002 *Phys. Lett. B* **537** 21
[9] Achasov N. N and Kiselev A. V, 2002 Phys. Rev. D **68** 014006
[10] The KLOE Collaboration, 2006 *Phys. Lett. B* **634** 148
[11] The KLOE Collaboration (Contributed paper to Lepton Photon), 2007 (arxiv:0707.4609)
[12] The KLOE Collaboration, 2009 *Phys. Lett. B* **681** 5
[13] Gokalp A, Kucukarslan A, Solmaz S and Yilmaz O 2008 *J. Phys. G.* **28** 2783

Journal of Physics: Conference Series **348** (2012) 012019 doi:10.1088/1742-6596/348/1/012019

The Study of Semileptonic B_c to D_{s1} Transitions in QCD Sum Rules

M Ghanaatian[1], R Khosravi[2] and A Bazrafshan[3]

[1]Physics Department, Payame Noor University, Jahrom, Iran
[2]Physics Department, Isfahan University of Technology, Isfahan, Iran
[3]Physics Department, Shiraz University, Shiraz, Iran

E-mail: m_ghanatian57@yahoo.com

Abstract. We consider the semileptonic Bc to D_{s1} transitions in the frame work of three-point QCD sum rules. Here, we investigate the D_{s1} meson as a conventional $c\bar{s}$ meson in the pure state. The obtained results for the form factors are used to evaluate the decay rates and branching ratios. Any future experimental measurement on these form factors as well as decay rates and branching fractions and their comparison with the obtained results in the present work can give considerable information about the structure of this meson.

1. Introduction

In this work, taking into account the gluon condensate corrections, we analyze the rare semileptonic $B_c \to D_{s1} l^+ l^-$, $l = \tau, \mu, e$ transition in three-point QCD sum rules (3PSR) approach. Note that, the $B_c \to (D^*, D_s^*, D_{s1}(2460))\nu\bar{\nu}$ transitions have been studied in Ref. [1], but assuming the D_{s1} only as $c\bar{s}$. The $B_c \to D_q l^+ l^-/\nu\bar{\nu}$ [2], $B_c \to D_q^* l^+ l^-$, (q=d,s) [3] transitions have also been analyzed in the same framework.

The rare semileptonic $B_c \to D_{s1} l^+ l^-$ decays occur at loop level by electroweak penguin and weak box diagrams in the standard model (SM) via the flavor changing neutral current (FCNC) transition of $b \to s l^+ l^-$. The FCNC decays of B_c meson are sensitive to new physics (NP) contributions to penguin operators. Therefore, the study of such FCNC transitions can improve the information about:

- The CP violation, T violation and polarization asymmetries in $b \to s$ penguin channels, that occur in weak interactions.
- New operators or operators that are subdominant in the SM.
- Establishing NP and flavor physics beyond the SM.

2. Transition form factors

To calculate the form factors within three-point QCD sum rules method, the following three-point correlation functions are used [1-4]:

$$\Pi_{\mu\nu}^{V-A}(p^2, p'^2 q^2) = i^2 \int d^4x\, d^4y e^{-ipx} e^{-ip'y} \left\langle 0 \middle| T[J_\nu^{D_{s1}}(y) J_\mu^{V-A}(0) J^{B_c^\dagger}(x)] \middle| 0 \right\rangle \quad (1)$$

$$\Pi_{\mu\nu}^{T-PT}(p^2, p'^2 q^2) = i^2 \int d^4x\, d^4y e^{-ipx} e^{-ip'y} \left\langle 0 \middle| T[J_\nu^{D_{s1}}(y) J_\mu^{T-PT}(0) J^{B_c^\dagger}(x)] \middle| 0 \right\rangle \quad (2)$$

where $J_\nu^{D_{s1}}(y) = \bar{c}\gamma_\nu\gamma_\mu s$ and $J^{B_c}(x) = \bar{c}\gamma_5 b$ are the interpolating currents of initial and final meson states, respectively. $J_\mu^{V-A} = \bar{s}\gamma_\mu(1-\gamma_5)b$ and $J_\mu^{T-PT} = \bar{s}\sigma_{\mu\nu}q^\nu(1+\gamma_5)b$ are the vector-axial vector and tensor-pseudo tensor parts of the transition currents. In QCD sum rules approach, we can obtain the correlation function of Eq. (1) and Eq. (2) in two sides. The phenomenological or physical part is calculated saturating the correlator by a tower of hadrons with the same quantum numbers as interpolating currents. The QCD or theoretical part, on the other side is obtained in terms of the quarks and gluons interacting in the QCD vacuum. We calculate the phenomenological part of the correlators in 3PSR. Therefore, we have,

$$\Pi_{\mu\nu}^{V-A}(p^2, p'^2 q^2) = -\frac{f_{B_c} m_{B_c}^2}{(m_b+m_c)} \frac{f_{D_{s1}} m_{D_{s1}}}{\left(p'^2 - m_{D_{s1}}^2\right)\left(p^2 - m_{B_c}^2\right)} \times [iA'^{B_c \to D_{s1}}_V(q^2)\varepsilon_{\mu\nu\alpha\beta} p^\alpha p'^\beta$$

$$+A'^{B_c\to D_{s1}}_0(q^2)g_{\mu\nu}+A'^{B_c\to D_{s1}}_1(q^2)P_\mu p_\nu+A'^{B_c\to D_{s1}}_2(q^2)q_\mu p_\nu] \quad (3)$$

$$\Pi^{V-A}_{\mu\nu}(p^2,p'^2q^2)=-\frac{f_{B_c}m^2_{B_c}}{(m_b+m_c)}\frac{f_{D_{s1}}m_{D_{s1}}}{(p'^2-m^2_{D_{s1}})(p^2-m^2_{B_c})}\times[iT'^{B_c\to D_{s1}}_V(q^2)\varepsilon_{\mu\nu\alpha\beta}p^\alpha p'^\beta$$
$$-iT'^{B_c\to D_{s1}}_0(q^2)g_{\mu\nu}-iT'^{B_c\to D_{s1}}_1(q^2)q_\mu p_\nu] \quad (4)$$

where A'_V, A'_0, A'_1, A'_2, T'_V, T'_0 and T'_1 are the transition form factors. On the QCD side, using the operator product expansion (OPE), we can obtain the correlation function in quark-gluon language. For this aim, by performing the double Borel transformations over the variables p^2 and p'^2 on the physical as well as perturbative parts of the correlation functions and equating the coefficients of the selected structures from both sides, the sum rules for the form factors $A'^{B_c\to D_{s1}}_i$ are obtained:

$$A'^{B_c\to D_{s1}}_i=-\frac{(m_b+m_c)}{f_{B_c}m^2_{B_c}f_{D_{s1}}m_{D_{s1}}}e^{\frac{m^2_{B_c}}{M_1^2}}e^{\frac{m^2_{D_{s1}}}{M_2^2}}\{-\frac{1}{4\pi^2}\int_{m_c^2}^{s'_0}ds'\int_{s_L}^{s_0}ds\rho_i^{V-A}(s,s',q^2)\,e^{\frac{-s}{M_1^2}}e^{\frac{-s'}{M_2^2}}$$
$$-iM_1^2M_2^2\langle\frac{\alpha_s}{\pi}G^2\rangle\frac{C_i^{V-A}}{6}\} \quad (5)$$

where $i=V,0,1,2$ and for the form factors $T'^{B_c\to D_{s1}}_j$, we get

$$T'^{B_c\to D_{s1}}_j=-\frac{(m_b+m_c)}{f_{B_c}m^2_{B_c}f_{D_{s1}}m_{D_{s1}}}e^{\frac{m^2_{B_c}}{M_1^2}}e^{\frac{m^2_{D_{s1}}}{M_2^2}}\{-\frac{1}{4\pi^2}\int_{m_c^2}^{s'_0}ds'\int_{s_L}^{s_0}ds\rho_j^{T-PT}(s,s',q^2)\,e^{\frac{-s}{M_1^2}}e^{\frac{-s'}{M_2^2}}$$
$$-iM_1^2M_2^2\langle\frac{\alpha_s}{\pi}G^2\rangle\frac{C_j^{T-PT}}{6}\} \quad (6)$$

where $j=V,0,1$. The s_0 and s'_0 are the continuum thresholds in B_c and D_{s1} channels, respectively and lower bound s_L in the integrals. We calculate the explicit expressions of the coefficients $C^{V-A(T-PT)}_{i(j)}$ correspond to gluon condensates.

3. Conclusion

we would like to consider the D_{s1} meson as the pure $|c\bar{s}\rangle$ state. We choose the values of the condensates, leptonic constants, CKM matrix element, quark and meson masses [5,6,7]. To calculate the branching ratios of the $B_c\to D_{s1}(2460[2536])l^+l^-$ decays, we use the total mean life time $\tau_{B_c}=(0.46\pm0.07)ps$ [7]. Our numerical analysis shows that the contribution of the non-perturbative part (the gluon condensate diagrams) is about 12% of the total and the main contribution comes from the perturbative part of the form factors. The values for the branching ratio of these decays are obtained as presented in Table 1, when only the short distance (SD) effects are considered.

Table 1. The branching ratios of the semileptonic $B_c\to D_{s1}(2460[2536])l^+l^-$ decays with SD effects.

MODS	BR	MODS	*BR*
$B_c\to D_{s1}(2460)e^+e^-$	$(5.40\pm1.70)\times10^{-6}$	$B_c\to D_{s1}(2536)e^+e^-$	$(2.91\pm0.93)\times10^{-6}$
$B_c\to D_{s1}(2460)\mu^+\mu^-$	$(2.27\pm0.95)\times10^{-6}$	$B_c\to D_{s1}(2536)\mu^+\mu^-$	$(1.96\pm0.63)\times10^{-6}$
$B_c\to D_{s1}(2460)\tau^+\tau^-$	$(1.42\pm0.45)\times10^{-8}$	$B_c\to D_{s1}(2460)\tau^+\tau^-$	$(0.68\pm0.21)\times10^{-8}$

References

[1] Azizi K, Khosravi R and Bashiry V 2008 *Eur. Phys. J.* C **56** 357
[2] Azizi K and Khosravi R 2008 *Phys. Rev.* D **78** 036005
[3] Azizi F, Falahati F, Bashiry V and Zebarjad S M 2008 *Phys. Rev.* D **77** 114024
[4] Ghahramany N, Khosravi R and Azizi K 2008 *Phys. Rev.* D **78** 116009
[5] Shifman M A, Vainshtein A I and Zakharov V I 1979 *Nucl. Phys.* B **147** 385
[6] Bashiry V and Azizi K 2007 *JHEP* **0707** 064
[7] Amsler C et al 2008 Particle Data Group *Phys. Lett.* B **667** 1

AUTHOR INDEX

Institute of Physics
Dirac House, Temple Back
Bristol BS1 6BE UK

CURRAN ASSOCIATES INC
proceedings
.com